The Unwritten Chronicles

Bo Yin

The Unwritten Chronicles

China's Ancestral Secrets Inscribed in Genes

Translated by Vicky Lianghui Sun

PETER LANG
New York · Berlin · Bruxelles · Chennai · Lausanne · Oxford

Bibliographic information published by the Deutsche Nationalbibliothek.
The German National Library lists this publication in the German National Bibliography; detailed bibliographic data is available on the Internet at http://dnb.d-nb.de.

Library of Congress Control Number: 2025040585

Cover image:
AdobeStock_1635177604

ISSN 2689-7970
ISBN 978-3-0343-6050-0 (Print)
ISBN 978-3-0343-6053-1 (ePDF)
ISBN 978-3-0343-6054-8 (ePUB)
DOI 10.3726/b23340

Published by Peter Lang Publishing Inc., New York (USA)

info@peterlang.com

This edition is an authorized translation from the Chinese language edition
Published by arrangement with CITIC Press Corporation

This publication has been peer reviewed.

www.peterlang.com

Contents

List of Figures

List of Table

Preface: Genes Know the Answer

Do the Peking Man of Zhoukoudian represent our Chinese direct lineage? If indeed, we stem from the African *Homo sapiens*, then why doesn't our complexion mirror their ebony shade? Which cultivated plants and beasts are native, and which are foreign, and how did they shape our ancestry? What are the origins of the multitude of ethnicities found within China, and how did they navigate the vast Chinese terrain? Do we share a common thread with ethnic groups scattered around the globe?

Genes know the answer.

Historians find themselves in a quandary, the farther they venture into the annals of time, the scarcer the historical records, rendering the true face of history elusive. This same predicament plagues archaeologists when they encounter sites bereft of written testimonies, leaving them bewildered as to which group or town these relics align with in the chronicles of yore. And yet, it is the field of paleoanthropology that most ardently laments, for it grapples with the prehistoric epoch, a time when words had not yet sprung forth, leaving only silent remnants such as ancient fossils and primitive tools.

Ironically, it is in the pre-written era that many profoundly influential events unfolded in human history. The evolution and migration of ancient humans over countless eons remain veiled in a wordless shroud, with but rare vestiges to afford us glimpses of the past. The domestication of major crops, the taming of fowl and livestock, and the genesis of navigation, pottery, metallurgy, and architecture all predate the advent of written language. As such, posterity often resorts to weaving tales to ascribe names to the pioneers of these innovations, deprived of solid evidence.

Even as history took its scriptural form, voluminous knowledge remains unrecorded, or worse, lost amidst the millennia-long flow of time. The infamous episode of "burning books and burying scholars"[1] serves as a quintessential example. Additionally, not every ancient chronicler bore the weight of responsibility. Some, driven by ulterior motives, manipulated and falsified history, exalting the victorious and disparaging the defeated.

Beyond the haze of unreliable literary accounts and the mute whispers of archaeological evidence, does no other avenue lead us to uncover further historical verity?

Since 1980s, a wondrous interplay of disciplines has graced the scientific stage. Molecular biology has crossed into the domains of paleoanthropology, history, and archaeology, harnessing ancient and contemporary human and animal genetics to unveil truths beyond the scope of conventional inquiry. Most notably, it has laid bare the intricate tapestry of prehistoric human migrations.

Why do genes reveal history? How do biologists achieve this?

In essence, genes bear the legacy of ancestry and the subtle hues of variation across generations. Through gene comparison, one can discern the genetic kinship between individuals and groups, gauging the proximity of ties or the stretch of distance to their common forebears. Thus, a web of connections forms among the contemporaries, and a thread links the ancient tribes to the present.

Consider the Y-chromosome in humans—its genes echo the fatherly line, passed down from sire to son, a familial torch carried forward. Similarly, within the mitochondria, maternally inherited genes unveil the maternal link, streaming from mother to daughter. From these Y-chromosomal whispers, we ascertain the

---

1 The episode commonly referred to as the "burning of books and burying of scholars" (Chinese: 焚书坑儒; pinyin: fénshū kēngrú), alternatively described as the incineration of classical texts and the execution of Confucian scholars, is traditionally attributed to Emperor Qin Shi Huang of the Qin Dynasty (221–207 BCE). According to historical accounts, in 213 BCE, a large number of philosophical and literary works were destroyed, and in 212 BCE, approximately 460 Confucian scholars were purportedly buried alive. These actions were allegedly undertaken to suppress intellectual dissent and consolidate the ideological dominance of Legalism as the official doctrine of the Qin state, at the expense of competing schools of thought such as Confucianism.

The historical veracity of these events, however, has been subject to considerable scholarly skepticism. The primary source for the narrative, Sima Qian's Records of the Grand Historian (Shiji), was compiled more than a century later during the Han Dynasty. Given Sima Qian's position as a court historian under a regime ideologically opposed to Qin policies, it is plausible that the account was shaped by political motives aimed at legitimizing the Han and discrediting their predecessors. Furthermore, subsequent Confucian scholars generally accepted the account without critical examination. As one modern historian has remarked, the moral conveyed by such narratives was clear: "If you take our life, Heaven will take the life of your dynasty."
—Translator's note

closeness of different male peers, revealing the ancestral bonds between antiquity's men and their modern counterparts, shedding light on the migration dance of history. This pattern extends to matrilineal mitochondrial genes. Moreover, since the domestication and dissemination of flora and fauna entwine with specific ancient societies and their wanderings, delving into the genetic realm of domesticated organisms unveils a trove of human history—a vivid account of antiquity's agriculture, pastoral pursuits, and nomadic odysseys.

Hence, molecular biology delves into history, and molecular biologists metamorphose into historians.

During my college days, an esteemed mentor uttered this truth: the termination of the "Cultural Revolution" (1966–1976) ushered back weathered geologists to academia. Alas, they found a transformed geological doctrine—plate tectonics reconstructed the very edifice of geological theory. Likewise, in the 1990s, as my university days blossomed, the "Out of Africa" theory, rooted in genetic exploration, graced lecture halls, leaving a seismic imprint on the annals of paleontology, archaeology, and history.

Molecular biology's "cross-disciplinary impact," akin to the seismic cataclysm of plate tectonics, remolds prehistoric lore, and the grand tapestry of human civilization. Previous scholars' comparative research on ancient human bones, stone relics, and clay pottery, though invaluable, craves a fresh evaluation—through the lens of molecular biology. Indeed, many prior findings may stand as bastions of truth, resilient and exact, yet a substantial corpus awaits correction.

In the annals of antiquity and the dust of archaeological remains, scientists of the molecular realm have unearthed an "unwritten chronicles." Deciphering the enigmatic genetic code inscribed upon it, they have illumined human history afresh. Ancestral origins, evolution's dance, migrations' meanderings, and the minutiae of daily existence have found new revelation.

Beyond genetic research, the analysis of trace elements and isotopes has also achieved "interdisciplinary" breakthroughs, entwining within the humanities. Humans and animals absorb specific elements from their environment and food while they are alive, leaving behind elemental information in their bones, teeth, and stems after death. Using this information, scientists can infer the environment they lived in, the food they ate, or the nutrients they absorbed during their time, all of which are parts of history.

The greatest advantage of using these high-tech methods, such as genetics and elements, to study ancient humans, archaeology, and history lies in their authenticity. Records penned by ancients, whimsically colored by omission, memory's haze, or biased designs—these do not define the scientific revelations uncovered by genes and elements. Here, science reigns supreme, built upon verifiable truths and objective data. The scrutineers persist, testing, replicating, and upholding the rigors of sound research, banishing the specter of deceit.

With decades of relentless pursuit, led by molecular biology, the high-tech arsenal emerges—a formidable force within the humanities, yielding countless insights. Its stores of wisdom burgeon, a testament to the narrative bound in the wordless historical record of our lineage. Behold, within its parchment, lies the unwritten chronicles—the genesis of species, the migration of tribes, the symphony of sustenance, and the cultural thread.

Now, the tome lies open—our unwritten chronicles beckons—a voyage of ancestral discovery, guided by the timeless wisdom inscribed within the genes.

CHAPTER 1

The Bitter Journey to Beatific Country: The Million Years of *Homo Erectus* in East Asia

In the year of 1929, amidst the melancholy of early winter, the desolate Dragon Bone Hill (龙骨山) bore witness to a scene of diligent excavation.

A group of individuals, under the auspices of China's Geological Survey and the Beijing Union Medical College, embarked on an archaeological quest within the environs of Zhoukoudian in Fangshan district. Their pursuit lay in unearthing the vestiges of prehistoric life concealed within the Quaternary strata. Among them, their leader, harboring a flicker of hope, could not help but imagine the profound impact that a discovery of ancient human remains would yield.

Suddenly, an intrepid member of the team stumbled upon a narrow orifice, no wider than a single person's passage. The incumbent guardian of the dig was none other than the esteemed archaeologist, Pei Wenzhong (裴文中), (1904–1982). Venturing within, he beheld a treasure trove of animal fossils. A revelation of immeasurable scientific import; thus inspired, he led his companions to persist in their toils amidst the biting cold.

As the sun dipped below the horizon on the 2nd of December 1929, and the world was enveloped in the shadow of dusk, the undaunted archaeological party illuminated their work with candlelight. It was at this very moment that Pei Wenzhong, aflutter with excitement, proclaimed, "Behold, a *hominid*!" Gathering around in unison, they beheld a portion of a hominid's cranial remains, exposed upon the ground.

Undeterred, their unwavering resolve knew no bounds, and they toiled throughout the night, until, at last, the entire hominid skull was laid bare. The fervor that gripped the archaeological company was undeniable, and as dawn broke, Pei Wenzhong dispatched a messenger to the hallowed halls of the Geological Survey Institute within Beijing's precincts, bearing tidings of their momentous find. Several days later, Pei Wenzhong, with utmost care, enshrouded the hominid skull in his own humble blanket, accompanying it on a carriage ride to the Beijing city.

Swiftly traversing the realms through the medium of telegraphic transmission, the news of Beijing's discovery reverberated throughout the globe: A relic of human antiquity, a fossilized cranium, lay unveiled in the heart of China!

"Could he/she be our Chinese ancestors?"

1.1 When Franz Weidenreich (1873–1948) met Peking Man

In the annals of archaeology, the fortuitous unearthing of ancient human skulls at Zhoukoudian was no mere stroke of luck.

Chinese people often sold the remains of ancient animals as traditional medicine, which had long attracted many foreign experts researching ancient animals and ancient humans. These experts found precious animal fossils in the so-called "dragon bones." In 1921, Swedish and American scientists came to Zhoukoudian, China, in search of prehistoric animal fossils. Guided by local quarry workers, they arrived at Dragon Bone Hill and found some uniquely shaped quartz flakes in the rock layers on the mountain. It was evident that these quartz flakes were not naturally formed but were shaped by human activity. This discovery was significant, and the scientists asserted that primitive humans must have lived here. Subsequently, they excavated some animal fossils from Dragon Bone Hill, took them back to a university in Sweden for research, and in 1926 announced that among these fossils were two primitive human molars.

This news excited one of the leaders of the archaeological work, the Canadian anatomist Davidson Black (1884–1934), who was then working at Peking Union Medical College in Beijing. He named the two teeth representative of the human species as "Sinanthropus pekinensis" (in Latin) or "Peking Man" for short. Soon, other scholars discovered that the Peking Man fossils were remarkably similar to those found earlier in Indonesia, known as Java Man. Consequently, they grouped the two together into the category "*Homo erectus*," which represented a long and important stage in human evolution.

However, the academic community remained cautious, as they felt that determining an important new species of ancient humans based on just a few teeth found in Zhoukoudian might lack sufficient evidence. To address this, Davidson Black secured funding from the Rockefeller Foundation in the United States and established an archaeological team to go to Zhoukoudian's Dragon Bone Hill. Luck was on their side, and the astonishing skull of Peking Man was discovered.

Since the first skull of Peking Man was found in 1929, Pei Wenzhong's assistant and paleoanthropologist Jia Lanpo (贾兰坡), (1908–2001) continued excavation work in Zhoukoudian and in November 1936, consecutively discovered three more Peking Man skulls. Through multiple archaeological excavations in the Zhoukoudian area, a total of six Peking Man skulls, 12 skull fragments, 15 mandibles, 157 teeth, and broken femurs from over 40 male, female, young, and old individuals were unearthed. Archaeologists also found around 100,000 pieces of stone tools and evidence of the use of fire, such as ash layers, burnt stones, and charred bones.

Alas, as the scholarly odyssey into the realm of Peking Man commenced, tragedy befell—Davidson Black succumbed to a sudden heart ailment in the spring of 1934, his untimely departure a somber enigma. It is said that on his desk lay the evocative skulls of both Peking Man and Java Man, enigmatically mirroring his life's pursuit.

Thus, perhaps destiny wove the first strands of adversity into the unfolding tapestry of Peking Man's journey.

In the spring of 1935, at the invitation of the Rockefeller Foundation, the German Jewish scientist, Franz Weidenreich, arrived in Beijing, weary from his journey. He was to replace Davidson Black and embark on a research journey concerning the Peking Man fossils. Weidenreich had previously served as a professor of anatomy at various renowned universities in Germany. However, due to his Jewish identity, he faced increasing threats from the Nazi regime in Germany. In 1934, he fled to the United States and eventually got in touch with the Rockefeller Foundation, who were seeking a scholar to succeed Davidson Black. Weidenreich's expertise in osteology, particularly in revealing the evolutionary processes of primates through skeletal anatomy, made him a perfect candidate. Thus, the fate of Franz Weidenreich intertwined with the Peking Man fossils.

Weidenreich conducted a series of studies on the Peking Man fossils, examining their mandibles, teeth, and skulls. He put forward many groundbreaking ideas, including the notion that Peking Man may have practiced cannibalism. Perhaps his most significant contribution to the study of the Peking Man fossils was that he "left a legacy." In 1941, as the storm clouds of World War II gathered, Weidenreich realized the urgency of the situation. Before returning to the United States, he had his assistant create a model of the Peking Man skull. As history unfolded, the original Peking Man fossils mysteriously disappeared during their transportation. Fortunately, the preserved skull model ensured that the research on Peking Man could continue.

Upon his return to the United States in 1941, Weidenreich continued his research on human evolution at the American Museum of Natural History. He studied various ancient human fossils, including Peking Man from Zhoukoudian and the Java Man from Indonesia. Weidenreich proposed the idea that human evolution from early humans to modern humans involved significant changes, such as bipedalism, increased brain capacity, and reduced facial size.

Now, the question arose: Could we, the modern humans of China, have evolved from Peking Man? In other words, were Peking Man our ancestors?

Weidenreich's answer was affirmative.

In 1943, he suggested that modern Chinese people were descendants of Peking Man. This conclusion was drawn from the many similarities between the Peking Man skull and those of modern East Asians, such as flat faces, high cheekbones, and shovel-shaped incisors. Three years later, he published the book "Apes, Giants, and Man," expanding this idea globally and proposing the "Multiregional Hypothesis" of human origins.

According to his hypothesis, modern humans did not all originate from a single source. Different ancient human populations across the world evolved independently and became distinct ethnic groups. Human evolution showed regional characteristics and could be divided into regions such as Australia, Mongolia, Africa, and Europe. One branch was the "System of the Giants," which evolved from ancient giants, through Java Man, Solo Man, Wadjak Man, and finally became the Indigenous Australians. Their evolution was unidirectional and unrelated to that of other regions' ancient humans. Another branch was the Chinese Man, which, based on the analysis of the Zhoukoudian fossils, displayed a series of early primitive features of the Mongoloid race. Thus, the Chinese Man discovered at Zhoukoudian should be considered the ancestor of the Mongoloid race, and modern Chinese people belong to this race. Additionally, the ancient human fossils found in Europe, such as the Neanderthals, would be the ancestors of present-day Europeans, while any ancient human fossils found in Africa would be the ancestors of modern Africans.

Furthermore, Weidenreich also believed that interbreeding between different human populations was possible during early human evolution. Therefore, the characteristics of Chinese Man might also appear in other modern human populations today.

Weidenreich's giant hypothesis has long been proven to be incorrect. However, his evolutionary speculation about the Zhoukoudian Peking Man—the multiregional hypothesis—is essentially a replica of the "continuity with hybridization" hypothesis proposed and upheld by some paleoanthropologists today.

Influenced by Weidenreich's theory, in the 1960s, American physical anthropologist Carleton S. Coon (1904–1981), systematically put forward the "multiregional origin" hypothesis in his book "The Origin of Races." This hypothesis suggests that all modern humans on Earth can be classified into five races: Caucasoids (European whites), Mongoloids (East Asians), Australoids (Australian Indigenous people), Negroids (African blacks), and Capoids (Khoisan people of Africa). According to this hypothesis, these races evolved separately with minimal gene flow between them, resulting in varying levels of civilization due to different rates of evolution.

At that time, very ancient human fossils had already been unearthed in Africa. Coon's explanation was that Africa was likely the birthplace of humans, but it was only the "nursery," while Eurasia served as the "school" for humanity. Early human ancestors left Africa and evolved into the five modern races. He used the analogy of a multibranched candlestick, commonly used by Europeans, to describe his hypothesis—where the five races are like five candlesticks that share the same base but quickly branch out from the roots, evolving into modern humans. Hence, the multiregional origin hypothesis is also referred to as the "candelabra model."

Coon's hypothesis contained elements of racial discrimination, and it faced strong criticism upon its introduction. Nevertheless, some of his scientific viewpoints on human origins and evolution received attention and even support from many scholars.

Subsequently, a group of paleoanthropologists, represented by Chinese scholars, further developed the scientific viewpoints of Weidenreich and Coon and proposed the "continuity with hybridization" hypothesis.

In simple terms, the continuity with hybridization hypothesis adds many lateral connections between the candles of Coon's "candelabrum," indicating varying degrees of gene flow between different races. The notion of "ancient humans from various regions separately evolving into modern humans," as proposed by Weidenreich and Coon, has gained affirmation and persistence from this group of scholars. They believe that modern humans in East Asia evolved from ancient humans in East Asia, while those in Europe and West Asia evolved from ancient humans in those regions, and modern humans in Africa descended from ancient African humans. We can colloquially refer to this group of scholars as the "fossilists."

Since the discovery of Peking Man in Zhoukoudian, many ancient human fossils and various types of stone tools used by ancient humans have been found all over the world. Paleontologists have found several key pieces of evidence in these discoveries to support their views. Among the most significant evidence are the shovel-shaped incisors and the evolution of stone tools.

What are shovel-shaped incisors?

Now, please use your tongue to lick the inner side of your two upper front teeth. Do you feel slightly protruding ridges on both sides of your front teeth? When your tongue slides from left to right on the inner side of your upper front teeth, do you feel like it's following a shovel-shaped path?

If your answer is yes, then you have shovel-shaped incisors, which means your upper front teeth are shaped like shovels, curled up on both sides and flat in the middle.

According to statistics, 80% to 90% of modern Chinese people have shovel-shaped incisors. This dental characteristic is very common among modern East Asians and Native Americans, with only about 10% of modern Europeans and around 15% of modern Africans having shovel-shaped incisors. Surprisingly, scholars have collected information on ancient human teeth unearthed from Chinese ancient tombs and found that, except for teeth found in Xinjiang Region of China, all other ancient human upper front teeth from different regions in China are shovel-shaped.

Even more astonishingly, paleontologists have discovered that in all periods of ancient human fossils found in China, whenever upper front teeth have been preserved, they have always been shovel-shaped incisors. Whether it's the 1.7-million-year-old Yuanmou Man from Yunnan province, the 700,000-year-old Peking Man from Zhoukoudian, or the tens of thousands of years old Hetao Man, they all have the same type of upper front teeth.

One scholar supporting the multiregional origin hypothesis lamented, "I really can't think of any factors in the living environment of China that would make it necessary for everyone living here to have shovel-shaped incisors. It seems that those without shovel-shaped incisors would have a hard time surviving and would be eliminated." Therefore, the characteristic of shovel-shaped incisors seems to have been passed down through generations of ancient humans in China and eventually inherited by modern Chinese people. In other words, the ancient humans in China underwent a long process of evolution to become modern Chinese people.

Another important piece of evidence from the paleontologists comes from the types of stone tools made and used by ancient humans.

Looking back in time, when early humans diverged from other primates to become a new species, their technological level was not much superior to their chimpanzee neighbors. The tools they could create were limited to simple tools made from stones and wood.

Around 3.3 million years ago, in what is now Kenya, East Africa, early humans created the earliest stone artifacts, which were simple stone cores and flakes used for cutting and smashing, making their hunting, gathering, and food processing activities more efficient. This marked the beginning of the Stone Age for early humans. Paleontologists call this period the Mode I, or, in technical terms, the Oldowan industry.

The Mode I underwent a long period of development for over one million years until about 1.7 million years ago, when early humans in East Africa finally experienced a significant technological innovation. Their stone tool-making techniques became slightly more sophisticated, enabling them to create various types of stone tools, particularly a hand axe with symmetrical sides, slender shape, and a sharp end on one side and a thick end on the other, which was a "must-have weapon for both home and travel, killing and plundering." Of course, at that time, these hand axes were primarily used for hunting wild animals rather than for harming each other. According to experts, this period belongs to the Mode II, and the technology early humans possessed is called the Acheulean industry.

After another long period of over one million years, around 500,000 years ago, early humans made further advancements and created more advanced stone tools, including a typical form of stone core with a prism shape, which was practical and easy to carry. There is still some debate in the academic community about the exact birthplace of these stone tools, with suggestions that they first appeared in South Africa, East Africa, or Western Asia. This period is referred to as the Mode III, and the professional term for this technology is the Levallois technique.

After summarizing the development history of stone tool technology over millions of years, we observe continuous innovation in stone tool technology in East Africa and its surrounding regions. However, in the vast prehistoric era, how did stone tool development unfold in East Asia?

Regrettably, it appears that stone tools in East Asia emerged as early as 1.7 million years ago, during a period dominated by the first-mode stone tools. While in East Africa, the second-mode and third-mode stone tools emerged subsequently, ancient humans in East Asia continued to use their first-mode cutting and smashing tools at a leisurely pace. Archaeologists conducted a study of stone tools and found that until approximately 60,000 years ago, there were hardly any traces of third-mode tools within China's borders. Throughout the Paleolithic era, or the era of stone tool production, East Asia, including China, seemed to be isolated from the rest of the world. Since the introduction of the first-mode stone tools 1.7 million years ago, stone tool technology developed independently without much technological exchange with the Western world for an extensive period.

Stone tool technology was created, inherited, and developed by ancient humans. So, what does the isolated development of stone tool technology in East Asia mean for ancient humans?

Paleontologists believe that this indicates the independent evolution of ancient humans in East Asia, with minimal or occasional interbreeding with ancient humans from other regions in the world. Those ancient humans in East Asia, who wielded stone tools to cut and smash plants and animals over a million years ago, are likely the ancestors of modern East Asians.

With supporting evidence from dental and stone tool evolution, along with other pieces of evidence, the viewpoint of the paleontologists seems to be firmly established as a factual account.

1.2 "Mitochondrial Eve" and "Y-Chromosome Adam"

However, as people in today's Internet era often say, "What defeats you is not your opponent, but cross-disciplinary challenges." The power that shook the traditional views of paleontology did not come from the field of paleoanthropology, nor from new dental or lithic evidence; it came unexpectedly from the realm of molecular biology.

In January 1987, a female doctoral student from the United States, Rebecca L. Cann (b. 1951), and her colleagues published a paper titled "Mitochondrial DNA and Human Evolution" in the prestigious British scientific journal *Nature*.[1] The main point of the paper was that there is only one place of origin for humans, likely in Africa, and the time of origin was within the last 200,000 years. According to their findings, all modern humans today share a common female ancestor.

At that time, this was a groundbreaking discovery, and it represented a disruptive "cross-disciplinary blow" to the field of molecular biology. If the conclusions of this paper were correct, the entire theoretical framework about human origins and evolution in paleoanthropology would have to be rewritten.

Let's take a closer look at how this "cross-disciplinary blow" occurred scientifically.

From the perspective of molecular biology, the human body contains about 100 trillion cells, each of which contains a cellular organelle called the mitochondrion, located in the cytoplasm outside the cell nucleus. The nutrients we ingest, such as sugars and fats, are ultimately oxidized in the mitochondria, releasing energy for the use of our body's cells. In simple terms, mitochondria are the "powerhouses" of our body's cells.

In Chinese high school textbooks, we learn that the genetic material of the human body is located on the chromosomes within the cell nucleus, where a vast number of genes are arranged in the double helix structure. However, mitochondria also contain a small amount of genetic material—mitochondrial DNA (mtDNA). Furthermore, we know that the beginning of a human life starts with the fusion of the father's sperm and the mother's egg cell. In other words, mitochondria,

---

1 Cann, R., Stoneking, M. & Wilson, A. "Mitochondrial DNA and human evolution," *Nature* 325, 31–36. (1987). <https://doi.org/10.1038/325031a0>.
—Translator's Note.

as cellular organelles, come from the mother's egg cell and not from the father's sperm (although a small amount of mtDNA is also present in the father's sperm, it is discarded during fertilization). Therefore, mtDNA is inherited from the mother's side. As a result, molecular biologists can compare people's mtDNA and analyze the maternal lineage, thus tracing their maternal ancestry.

Rebecca L. Cann collected 147 samples of mtDNA from different populations around the world, using human placenta as the material source because it contains abundant mtDNA. She discovered that by tracing the maternal lineage of all the sample owners through their mtDNA, they all converged to a single female ancestor, regardless of where the individuals with the mtDNA lived in the world; they were all descendants of that female ancestor.

In a time when media drew inspiration from the earliest human female mentioned in the Bible, Eve, they affectionately dubbed her "Mitochondrial Eve." Within the genetic fabric, Rebecca L. Cann unearthed a trove of revelations about this female ancestor.

The enigma lingered: where did Mitochondrial Eve once tread upon Earth's vast canvas?

Picture a tranquil pond, as a stone gracefully breaks its placid surface. With each ripple encircling outward, we find ourselves privy to an artful revelation. Though the stone's descent eludes us, the center of the ripples divulges its hidden abode. Much akin to this delicate dance, deciphering the dwelling of Mitochondrial Eve mirrors the discernment of the stone's landing spot through its ethereal ripples.

Delving deeper, we discover that with each generation, human genes dance with rare serendipity. Occasional genetic mutations occur, bequeathing subtle dissimilarities from their forebears. In essence, the chasm between progeny and progenitor, the rings of time, widens, nurturing a symphony of genetic diversity. Assuming an ancient era, where each human generation's lifespan bore resemblance, we glean a rudimentary truth: the farther temporal shores, the grander the genetic disparities between outermost "ripples" and the heart of the "ripple."

Such is the course of mtDNA inheritance. From mother to child, a timeless ritual unfolds, with whispers of genes marked by cosmic brushstrokes of mutation. Hence, progeny don their individual array of genes, a vast array of human biodiversity. Drawing parallels among modern populations, molecular biologists unveil not only the bonds that bind diverse human groups but also unravel the tapestry of their female ancestors' existence and origins.

Rebecca L. Cann's inquiry revealed that diversity's brush paints with unique strokes among modern human groups. With Africa as the epicenter of genetic opulence, followed by Asia and Europe, the genes of Native American brethren swayed less melodiously. In this symphony of diversity, whispers from antiquity resound, divulging a shared ancestress among modern humans. A common female forebear whose ancient echoes resonate from where the hues of genetic diversity flourish—the African continent, where the stone of genesis first graced the ethereal waters.

Then, another profound question lingers on the wind: during which epoch did our matriarch, Mitochondrial Eve, grace the stage of history?

Before unfurling this ageless enigma, we invite you, dear reader, to contemplate: did she adorn the Earth's stage some 5 million years ago, or did her presence linger closer, at 1 million, 500,000, or perhaps a mere 100,000 years ago? Where lies the elusive whisper of truth?

While genetic mutations within human essence are rare, time's elegant brushstrokes whisper of a constant tempo. On average, 30 genes sway in dance to a gentle mutation in each successive generation, uniting the past and present in graceful continuity. A dance of genes, fewer in number than those encased within the cell's sacred nucleus. Armed with the knowledge of the tempo of these genetic waltzes, and delicately estimating the subtle cadence between each generation, the veil of time is unveiled, revealing the era when Mitochondrial Eve walked the sacred threads of life.

Rebecca Cann's manuscript whispers the truth: a symphony of human history echoes from 140,000 years ago, a haunting serenade woven into the very fabric of our existence.

All of us, modern beings, trace our lineage back to a female ancestor who walked the African lands some 140,000 years ago.

Let us not mistake this to suggest that she was the sole female inhabitant of Earth in that distant era. No, certainly not. Many other women graced our planet at that time, each serving as confidantes to our Mitochondrial Eve. Regrettably, the descendants of these companions failed to persist through the ages until our present day.

To illustrate further, envision a quaint village nestled near the heart of the Chinese Tang Dynasty (618–907), around 1,400 years ago. Within its confines dwelled ten families, each possessing a distinctive recipe for lamb stew, a treasured secret passed down exclusively from mother to daughter. As the centuries unfurled, some families, alas, bore only sons, thereby losing the precious knowledge of their cherished culinary heritage. In the end, only one lineage endured, consistently birthing daughters who dutifully safeguarded and handed down the age-old recipe. Even as these daughters wed and established new households, they brought with them the coveted family legacy, entrusting it to their own progeny. Over time, subtle variations might have graced the recipe, dictated by individual palates or the ebb and flow of ingredients, but the essence of the original remained intact.

In this allegory, the woman of the Tang Dynasty village embodies our Mitochondrial Eve, the guardian of the precious recipe. The formula, in turn, represents mtDNA, and the slight adjustments signify genetic mutations.

Though we have now discovered Mitochondrial Eve, the quest to find our "Adam"—the common male ancestor of all modern humans—beckons with a touch more complexity. Again, as Chinese high textbooks teach, human sex is determined by sex chromosomes. Females carry two X chromosomes, while males possess one X and one Y chromosome. MtDNA gracefully passes through the maternal line, while the Y chromosome cradles the genetic inheritance exclusively transmitted from father to son. Thus, to uncover the shared male ancestor, the focus must turn to the genes residing on the Y chromosome. Alas, this journey proves more arduous, as the vastness of the Y chromosome's genetic content presents formidable challenges, especially with the limitations of the twentieth-century biological technology.

It was only in the year 2000, a full 13 years after unearthing Mitochondrial Eve, that scientists finally unearthed the "Y-Chromosome Adam." Like his mitochondrial counterpart, this remarkable common male ancestor resided in Africa. His genetic affinities closely intertwined with a group known as the Khoisan, suggesting that he once thrived amidst their ancient community some 60,000 years ago.

Parallel to the tale of Mitochondrial Eve, the era of Y-Chromosome Adam was replete with other male companions who lived their lives. Yet, alas, they failed to leave a male legacy for the present age.

Now, let's return to the topic of the Peking Man at Zhoukoudian. Based on the discoveries in molecular biology mentioned earlier, a weighty conclusion can be drawn: the Peking Man at Zhoukoudian is not the ancestor of modern Chinese people.

How can this be explained? The era in which the Peking Man lived is far more ancient than that of Mitochondrial Eve and Y-Chromosome Adam. To determine the age of the Peking Man, scientists initially studied the stratigraphy and the fossils contained within the layers of the Earth. If we observe steep cliffs in the wild mountains or along riverbanks, we will notice that rocks and soil are layered on top of each other. This is because different layers of rocks and soil were formed and deposited at different times. A general rule in geology is that older layers are buried below, while more recent layers are distributed above. This is known as the "law of superposition" in geology. Different kinds of plants and animals lived during different geological periods, and their remains would be buried and preserved together with the rocks and soil of their time. Generally, the remains of ancient plants and animals are found in older layers, while those of more recent plants and animals are found in younger layers. This rule is referred to as the "law of faunal succession." These two principles are important means of determining the age of living beings, including ancient human beings.

In the late 1970s, scientists carefully studied the distribution of strata and the contained fossils at the Zhoukoudian Peking Man site. In the same era and region where the Peking Man lived, various other ancient organisms coexisted, such as ancient Smilodon (saber-toothed tiger), Chinese badger, Zhoukoudian rhinoceros, Sanmen horse, Jushi beaver, as well as the relatively less ancient Naramagou long-toothed elephant (*palaeoloxodon namadicus* in Latin), Yan Mountain rhinoceros, Sinomegaceros, cave bear, hyena sinensis, and more recent species such as wolves, brown bears, and numerous small rodents. Based on this information, scientists determined that the era in which the Peking Man lived was approximately 500,000 to 400,000 years ago.

By the twenty-first century, scientists had access to more advanced dating methods, using isotopic content in chemistry for dating purposes. Isotopes refer to elements that have the same number of protons but different numbers of neutrons in their atomic nuclei, making them isotopes of each other. Some isotopes are radioactive, and after an animal dies, certain isotopes within its body will steadily transform into other isotopes. Therefore, scientists can infer the age of an animal's life based on the proportion of isotopes found in its remains.

Researchers collected rock and stone tool samples from the Zhoukoudian Peking Man site and conducted repeated isotope dating. The conclusion was that the earliest period of Peking Man's existence dates back to approximately 770,000 years ago, with a margin of error of 80,000 years. Around 700,000 years ago, the Earth was experiencing a relatively cold period, which led to a higher evaluation of Peking Man's adaptation to cold environments and their ability to create fire artificially.

The latest traces of ancient human activity found at the Zhoukoudian Peking Man site were dated to approximately 230,000 years ago. In other words, after that period, the Peking Man disappeared from Zhoukoudian forever.

The Peking Man lived from approximately 770,000 to 230,000 years ago, while the common female ancestor of modern humans lived around 140,000 years ago, and the common male ancestor lived around 60,000 years ago, with no overlapping timeframes. Based on this chronological data, the Peking Man cannot be the ancestor of modern humans, as their descendants did not survive to the present day.

The molecular biology perspective on human evolution, known as the "Out of Africa" hypothesis, dauntlessly challenges the multiregional origin hypothesis and the subsequent "continuity with hybridization" hypothesis. In this intellectual clash, proponents of the former, affectionately dubbed as the "Geneticists," find themselves at odds with the paleoanthropologists. This scholarly dispute, like a dance of words, continues to captivate minds from diverse scientific domains.

For molecular biologists to eloquently refute the multiregional origin or continuity with hybridization arguments put forth by their paleoanthropological counterparts, they must offer thoughtful explanations for the compelling evidence presented. Particularly, aspects such as the distinctive shovel-shaped incisors and the enthralling evolution of stone tools demand careful examination and thoughtful discourse. On the other hand, stalwart paleoanthropologists, champions of the multiregional origin or continuity with hybridization hypotheses, must artfully challenge the conclusions drawn from DNA research by their molecular biologist colleagues.

As we embark on this intellectual odyssey, our quest aims to unveil the tapestry of Chinese civilization from a genetic perspective, delving into the enigmatic tales of our forebears. Thus, our focus shall gracefully rest upon the elucidation of molecular biology's counterarguments against the tenets of the paleoanthropologists' hypotheses. However, before proceeding with our spirited repartee, it behooves us to traverse the grand epochs of human evolution, understanding the precise stages that engender the scholastic debate.

We traverse back to the genesis of ancient humans, borne of venerable African forest apes some seven to five million years ago. These early progenitors, graced with brain capacities of approximately 500 milliliters, first awkwardly endeavored to stand on two feet, evoking the critical demarcation between humans and their primate kin. Amidst the tapestry of ancient Africa, perhaps multiple strains of Southern apes thrived, among which could reside our direct ancestral lineage.

Around 3 to 2.5 million years ago, a "new" ancient human with a brain capacity of around 800 milliliters appeared, named *Homo habilis* because they were able to make and use the earliest stone tools. The most famous *Homo habilis* fossil is "Lucy," the skeleton of an adult female found in Ethiopia in 1974. Paleoanthropologists named her "Lucy" after the popular song by The Beatles at that time. The first mode of stone tools made by ancient humans mentioned earlier was most likely invented and produced by *Homo habilis*.

As ancient humans' stone tool technology became more refined and the proportion of high-energy animal food in their diet increased, around 2 million years ago, *Homo habilis* evolved into *Homo erectus*.

Although *Homo erectus* descended from *Homo habilis*, there was a significant difference in brain capacity and body size between the two. Compared to *Homo habilis*, some *Homo erectus* individuals had a brain capacity that was 50% larger and were nearly 30 centimeters taller. Relying on their two feet, *Homo erectus* could run, and when a group of them acted collectively, they could even hunt large animals.

At the *Homo erectus* stage, ancient humans could finally travel long distances and leave their homeland, Africa, which they had inhabited for millions of years. Indeed, shortly after their appearance, they were found in the adjacent western part of Asia. In other words, apart from Africa, there is no trace of ancient humans on any other continent earlier than 2 million years ago, only ancient apes and monkeys.

Beyond the frontiers of Africa, the captivating tale of our ancient past unfolds in Georgia in Europe, where the earliest known *Homo erectus* beyond the African plains etched their mark in history. Endowed with the name "*Homo georgicus*," these early *Homo erectus* bearers, akin to *Homo habilis* in stature at about 150 centimeters, boasted cranial capacities that eclipsed their contemporaneous *Homo habilis* brethren. Wandering through epochs, they graced our annals with their adept usage of fire, making *Homo georgicus* the earliest known ancient humans to harness its elemental power.

Venturing forward in time, our gaze alights upon a momentous event some 700,000 years in the past as *Homo erectus* ventured to the hallowed grounds of Zhoukoudian. In their late *Homo erectus* stage, their cranial capacity extended to approximately 1,100 milliliters, akin to modern humans but shy of the soaring heights of *Homo habilis*' 1,350 milliliters. The bucolic embrace of Zhoukoudian, nestled amidst the Western Hills' craggy grandeur and the fertile expanse of the North China Plain, enticed them to linger, embracing the bounties of nature for countless millennia.

The prevailing consensus among esteemed paleontologists and distinguished geneticists revolves around the notion that *Homo erectus*, as the ancient human predecessors, ventured forth from the African cradle into the enchanting realms of East Asia. This harmonious belief between the two scientific realms resonates with utmost certainty.

Yet, the saga of human evolution neither ceased nor relented. Within the temporal confines of 300,000 to 200,000 years ago, an awe-inspiring transformation unfurled among the ancient Homo lineages—an exponential expansion in cerebral capacity, a meteoric rise in cognitive prowess. The *Homo erectus* embarked on a journey, metamorphosing into *Homo sapiens*, more aptly dubbed "wise humans." This erudite species, surpassing their antecedents, showcased an unprecedented flair for crafting intricate tools, communing through a tapestry of language, and portraying profound behavioral intricacies. Their hunting acumen transcended horizons. We, the modern denizens of this planet, embody the legacy of *Homo sapiens*—creatures of intellect and ingenuity, united in this shared mantle of sapience.

The divide among scholars emanates not from the heredity of *Homo sapiens*, which traces its origins to *Homo erectus*, but rather from the eternally enthralling question—was the *Homo sapiens* diaspora a grand exodus, unfurling from Africa's bosom to distant lands (Figure 1.1), or rather, a mosaic of evolutionary tales, with diverse regions nurturing the evolution of *Homo erectus* into *Homo sapiens* independently (Figure 1.2)?

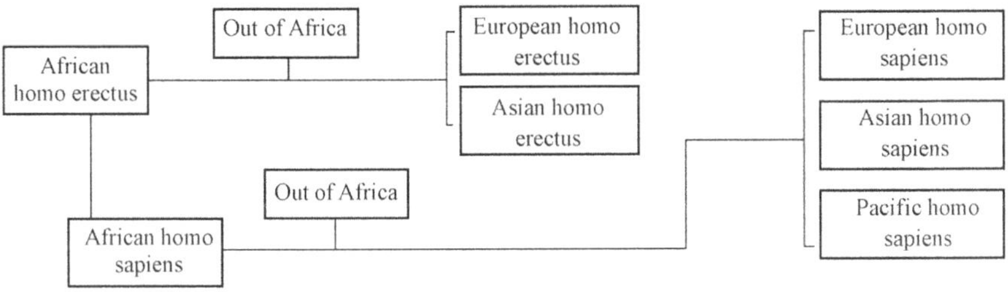

Figure 1.1: The Out of Africa hypothesis.

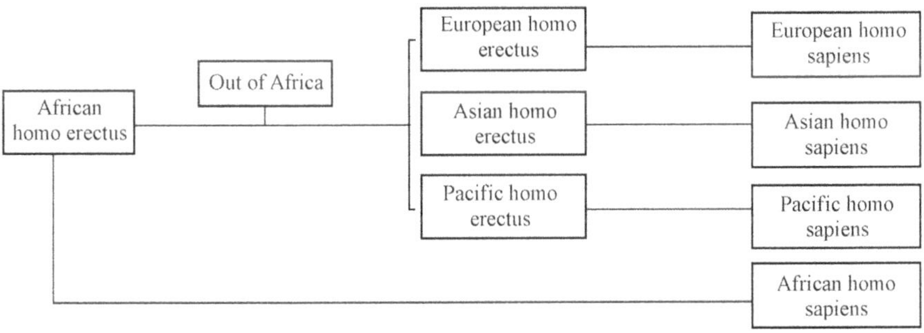

Figure 1.2: The multiregional origins hypothesis.

Out of Africa

Homo erectus africanus, *Homo erectus* europaeus, *Homo erectus* asiaticus, *Homo erectus* oceanicus

Homo sapiens africanus, *Homo sapiens* europaeus, *Homo sapiens* asiaticus, *Homo sapiens* oceanicus

1.3 Incisors, Stone Tools, and the Enigma of "Neighbor's Lost Axe"

Within the realm of geneticists' beliefs lies the notion that all modern human beings, regardless of their geographic origins, trace their ancestry back to a particular kind of *Homo sapiens* that emerged from the evolutionary path of *Homo erectus* in Africa. This stands in contrast to the idea that various *Homo erectus* populations independently evolved in distinct regions. Hence, the lineage of present-day East Asians does not stem from the evolution of Peking Man at Zhoukoudian but rather stretches back to the distant African continent. Nevertheless, a puzzling paradox emerges: how do the proponents of the genetic theory reconcile the presence of shovel-shaped incisors among ancient East Asian hominins from disparate periods, a characteristic seemingly at odds with their perspective?

To unravel the enigmatic origins of these shovel-shaped incisors, one must journey back to the era of Franz Weidenreich and even trace the footsteps of his forerunner, the eminent German naturalist Ernst Haeckel (1834–1919).

In Germany, while Haeckel, advocated the evolutionary principles of famed Englishman Charles Darwin ((1809–1882), he nonetheless diverged from the notion of African origins for humanity. Instead, he firmly upheld the belief that the orangutans of Southeast Asia represented the closest living relatives to humans, proposing that Asia, home to these apes, could be the cradle of humanity. In reality, during that era, some individuals had already dissected both the Asian orangutans and the African chimpanzees, finding that the latter's body structure was more similar to humans. But Haeckel was stubborn in his views, likely due to being influenced by his religious beliefs and political tendencies.

During that period, certain German scholars propagated the concept of an "Aryan race," claiming that the Aryans emanated from Northern India and were a noble lineage that had long ago conquered Europe. They contended that the Germanic people were direct descendants of the Aryans. Consequently, Haeckel and his compatriot naturalists, swayed by a sense of perceived superiority, subscribed to the idea that the birthplace of humanity rested in Asia, rather than elsewhere. Haeckel envisioned a lost continent "Lemuria" as humanity's initial abode, which had since submerged beneath the depths of the Indian Ocean, rendering the search for early human fossils an arduous endeavor.

Inspired by Haeckel's teachings, a surge of individuals set forth on journeys to Asia in pursuit of ancient human remains. Among them, a doctor from the Dutch East India Company ardently embraced Haeckel's supposition, organizing an expedition to unearth ancient human fossils in Indonesia. Serendipitously, in 1891, he stumbled upon a significant discovery on Java Island—ancient human cranial fragments and a femur, now renowned as the "Java Man." This fortuitous find further elevated Haeckel's stature within the scientific community.

Franz Weidenreich, while unlikely to have endorsed racist ideologies due to his Jewish heritage, could not entirely escape Haeckel's intellectual influence within the German academic sphere, where Asia was perceived as humanity's cradle. Consequently, when he arrived in Beijing and encountered the Peking Man fossils at Zhoukoudian, he unconsciously imbued them with ancestral significance for modern Chinese people. Initially, he also believed that Asia was the birthplace of humanity. With this mindset, when he arrived in Beijing and faced the Zhoukoudian Peking Man fossils, he subconsciously assumed they were the ancient ancestors of modern Chinese people and began searching for connections between the fossils and modern Chinese individuals. Already aware that shovel-shaped incisors were a distinctive feature of East Asian populations, Weidenreich went on to identify the teeth of the Zhoukoudian Peking Man as having shovel-shaped incisors as well.

Clearly, Weidenreich fell into a cognitive bias and, to make a less precise analogy, committed an error akin to the fable in the ancient Chinese book "Liezi" about "Suspecting One's Neighbor of Stealing the Axe."

In that story, a person lost his axe and suspected the neighbor's son of stealing it. He believed that the child, in speech, behavior, and appearance, seemed to fit the profile of a thief. However, he later found his lost axe in a valley and, upon returning to look at the child again, he no longer seemed like a thief. The change was not in the child, but in the psychological perception of the axe owner. Weidenreich's identification of the Zhoukoudian Peking Man was likely influenced by a similar psychological confirmation bias.[2]

In reality, all human incisors, including those in the upper jaw, serve the function of tearing and cutting food, resulting in some similarities among them. The confirmation bias led Weidenreich astray, causing him to misidentify the upper jaw teeth of the Zhoukoudian Peking Man as having shovel-shaped incisors.

The divergent thinking of Haeckel and Weidenreich has even affected scholars' identification of later-discovered Chinese ancient human fossils, such as the Yuanmou Man fossil found in Yunnan province.

In 1965, geologists discovered some mammal fossils and two ancient human teeth in Yuanmou, Yunnan. In 1973, paleoanthropologists identified and published a paper on the teeth. The paper identified the two teeth as upper lateral incisors, describing them as "shovel-shaped, large, and flat." They bore some similarities to the teeth of Peking Man from Zhoukoudian, but there were also differences.

---

2 The psychological term that could describe the man's behavior in this Chinese story is "confirmation bias." Confirmation bias, a phrase coined by English psychologist Peter Cathcart Wason (1924–2003), is the tendency of people to favor information that confirms or strengthens their beliefs or values and is difficult to dislodge once affirmed.

Wason, Peter C. (1960), "On the failure to eliminate hypotheses in a conceptual task," *Quarterly Journal of Experimental Psychology*, 12 (3): 129–140, <https://doi.org/10.1080/17470216008416717>

—Translator's note.

Based on these two teeth, Yuanmou Man was classified as an early human, and considering the geological layers, dated back to around 1.7 million years ago, making it the oldest ancient human found in China. Consequently, it gained a place of honor in textbooks for primary and secondary education, becoming well-known among the Chinese populace.

However, there have been disputes within the scientific community regarding this conclusion, far from being as "conclusive" as portrayed in textbooks. First and foremost, the dating of Yuanmou Man is a matter of contention.

The Yunnan region experiences significant rainfall, and its loose geological layers suggest that the age of the layers where the two teeth were found might not correspond directly to the time when Yuanmou Man lived. The teeth could have easily shifted position over time. A Chinese expert in loess research once used paleomagnetic techniques to date the geological layers of Yuanmou Man and proposed an age of around 600,000 to 500,000 years ago, a view supported by many scholars. As mentioned earlier, *Homo erectus* likely originated in Africa around 2 million years ago, and the earliest representative in Western Asia, Georgia Man, lived around 1.8 million years ago. Yunnan is not only distant from Africa but also vastly different in natural environment from the early *Homo erectus* habitats. It is challenging to imagine that *Homo erectus* appeared in Africa and then migrated to East Asia approximately 100,000 years later, completely changing their way of life.

Besides, there are doubts about the teeth of Yuanmou Man. First, basing the identification of an ancient human on just two teeth lacks robust scientific evidence; paleoanthropologists typically favor skull fossil evidence. Second, it is a stretch to claim that the two teeth are spade-shaped incisors. Without the influence of the concept of spade-shaped teeth, it is difficult for an ordinary person to identify the teeth as spade-shaped or even as incisors. The two teeth are more akin to the premolars of modern humans. Furthermore, the lingual aspect of the teeth is different from that of modern humans' spade-shaped incisors, as they exhibit pronounced longitudinal ridges in the middle of each tooth.

Regrettably, due to the influence of the confirmation bias, some scholars have insisted on identifying the teeth of Yuanmou Man as spade-shaped incisors. Consequently, the hypothesis that East Asian modern humans evolved from Yuanmou Man and Peking Man is put forth as a natural consequence.

Let us consider another intriguing tooth identification story. The earliest identified ancient human fossils in China were teeth and limb bones discovered in 1922 by French paleoanthropologists on the banks of the Salawusu River in the Ordos region of Inner Mongolia, named "Hetao Man." Scholars have proposed various dates for the time Hetao Man lived, ranging from 50,000 to 30,000 years ago and 140,000 to 70,000 years ago. Based on the tooth shape, Hetao Man's upper incisors were confidently identified as spade-shaped incisors, making them one of the pieces of evidence for the fossilists.

However, in the twenty-first century, a segment of Hetao Man's femur, collected many years ago, was sent to high-tech laboratories in both the United States and China for dating. The results from both laboratories were consistent, dating the femur to around 300 to 200 years ago. Yes, we didn't miss a zero; the femur belonged to a person from the Qing Dynasty, not an ancient human. Another set of limb bones found at the site was sent to a laboratory in Germany, dating back over 2,700 years ago, belonging to a person from the Spring and Autumn Period (771–256 BCE).

So, they all belonged to modern humans from the East Asian region, not ancient humans, and, of course, they had spade-shaped incisors.

In conclusion, according to the geneticists' perspective, the ancient human teeth found in China that predate *Homo sapiens* do not possess spade-shaped incisors and cannot prove that modern Chinese humans evolved from those ancient humans.

So, why do modern Chinese people have shovel-shaped incisors that are rare among other populations worldwide? The explanation lies in genetics.

Molecular biologists have identified a gene mutation in the human body that determines the development of shovel-shaped incisors. They have estimated that this gene mutation occurred approximately 30,000 years ago during the *Homo sapiens* stage, as earlier hominids such as Neanderthals did not possess this mutation.

Based on genetic comparisons of different populations around the world, molecular biologists speculate that the location of the gene mutation may have been in the central region of China. Subsequently, populations carrying this gene gradually spread to East Asia, leading to high proportions of individuals with shovel-shaped incisors in countries like China, Japan, and Korea. As ancient humans migrated to the Americas, this mutation also spread there, resulting in high proportions of shovel-shaped incisors among indigenous peoples.

According to the principles of evolution, a gene that spreads widely in a population and becomes highly prevalent usually signifies that it provides significant advantages for survival and reproduction, leading individuals with that gene to leave more offspring.

However, the question arises as to how much of a survival advantage shovel-shaped incisors provided to ancient humans. It seems unlikely that ancient humans gained more mating opportunities and higher reproductive success solely due to their dental aesthetics. So why do populations outside East Asia and the Americas lack this gene mutation and the corresponding dental feature?

The actual reason is that this gene mutation not only determines the development of shovel-shaped incisors but also leads to increased sweat glands, sebaceous glands, and mammary glands in carriers. In other words, it made carriers more "greasy." Many genes in our bodies have multiple functions, and a single gene can influence various characteristics of the human body.

In today's Chinese society, the notion of "greasy man" (*youni dashu*, middle-aged men who have aged not so gracefully) may be looked down upon, but for the ancient humans in East Asia thousands of years ago, the new gene mutation to be "greasy" was advantageous in helping them sweat and cool down in hot environments. The climate in the central and southern regions of ancient China was once very hot. Thus, populations carrying this gene mutation gained a survival advantage in East Asia, and over time, this gene became prevalent among East Asian populations. It is worth noting that the sweat and cooling theory is still speculative, and scientists have not yet reached a definitive conclusion. Nevertheless, they are certain that this gene mutation must have provided significant survival advantages to East Asian populations 30,000 years ago.

In summary, shovel-shaped incisors did not provide any specific survival advantage, nor did they impose any survival disadvantage. This dental characteristic merely hitched a ride on the gene mutation's coattails, following advantageous traits through generations.

Regarding another important piece of evidence for the fossil theory—the independent and continuous evolution of East Asian stone tool technology—molecular biologists cannot refute it with genetics. After all, stone tools are not biological entities; they do not have genes. For the evolution of stone tools, the fossils and the technological development they exhibit are there for all to see, and there is no dispute about them. However, when it comes to the relationship between technological evolution and human evolution, interpretations vary among different scholars.

The divergence between the two schools of thought lies in whether there is a technological progression from the stone tools of the *Homo erectus* found in East Asia to those of *Homo sapiens*. If there is a technological progression, does it prove that the local *Homo erectus* evolved into *Homo sapiens*?

The first question has a taste of the "Suspecting One's Neighbor of Stealing the Axe."

For ancient humans, when they migrated to distant lands, they would not carry large and cumbersome stone tools. At most, they would carry some small and lightweight ones. The process of human migration was lengthy and often spanned many generations. Therefore, it is natural that they would lose some stone tool-making techniques during the migration and, of course, improve some others. As a result, there would be differences in stone tool technologies in different parts of the world. For the East Asian region located at the eastern end of the Eurasian continent, with mountains, plateaus, deserts, and wastelands blocking their way, geographical barriers were significant. As a result, stone tool technologies developed relatively independently in this area, showing distinct differences from other regions, which is reasonable.

Additionally, after ancient humans arrived in a new environment and settled there for a long time, they would adapt to the local conditions and modify their original stone tool-making techniques based on the types of local rocks and available food resources. This resulted in the creation of stone tools suitable for local use.

Whether it was *Homo erectus* or *Homo sapiens*, they had to cross mountains and rivers, travel thousands of miles to reach East Asia, and then face similar East Asian environments. The loss and improvement of technologies, as well as the adaptation of technologies to local conditions, would occur in each era. Imagine this: even if the distant *Homo sapiens* initially had stone tool-making techniques vastly different from those of the East Asian *Homo erectus*, by the time they traveled across thousands of mountains and rivers to reach East Asia, their technology would have already undergone changes from the starting point or midway through the journey. It is even possible that they experienced technological regression due to significant technology loss. By comparing different stone tool features, people would easily find some similarities between the stone tools of these *Homo sapiens* and those of the local *Homo erectus*. But is this similarity due to actual inheritance of stone tool-making techniques, or is it merely an illusion that becomes more apparent the more we look, just like the confirmation bias in the Chinese fable "lost axe of the neighbor?"

It is hard to say.

Regarding the second question, scholars from the genetics perspective have made their point clear: "Without ancient genetic evidence, we cannot prove that East Asian *Homo erectus* left descendants."

This is truly a decisive move and makes sense. In the end, we can only prove direct kinship between an ancestor and a descendant through genetic comparison. The geneticists have presented evidence showing that all modern humans around the world are descendants of Mitochondrial Eve and Y-Chromosome Adam, who lived in Africa around 200,000 years ago. This evidence negates the view that Asian *Homo erectus* or European *Homo erectus* evolved into modern humans. The paleontologists' judgment that the local *Homo erectus* evolved into *Homo sapiens* based on the progression of stone tool technology is merely a conjecture.

Therefore, although there is still controversy surrounding the evolution of ancient humans in East Asia, scholars worldwide tend to believe that modern humans in East Asia did not evolve from East Asian *Homo erectus*. Consequently, the Peking Man of Zhoukoudian is not our direct ancestor; our direct ancestors should be the modern *Homo sapiens* who migrated from Africa, and if we continue to trace back, it should be the African *Homo erectus* that evolved into modern *Homo sapiens*.

1.4 East Asia's Version of the "Tasmanian Technological Tragedy"

Now, perhaps we should ask the question: "Why didn't the Peking Man leave descendants until today?" They once inhabited the Ape Man Cave on Dragon Bone Hill for 500,000 years, which proves that they were a group of people adapted to the local environment. Why did they permanently leave Dragon Bone Hill and disappear over 200,000 years ago?

Franz Weidenreich once put forward a speculation based on the features of the Peking Man's skull, suggesting that cannibalism existed within these groups of *Homo erectus*, similar to some modern-day tribes in the jungle. Some have inferred from this speculation that the extinction of the Peking Man might be related to self-inflicted violence. However, scholars have long dismissed Weidenreich's "cannibalism" hypothesis and, in turn, rejected the idea of the Peking Man's extinction due to self-inflicted violence.

A popular explanation is that the Peking Man encountered a harsh Ice Age, struggling to stay warm, and faced food scarcity in their living environment, ultimately succumbing to hunger and cold. This explanation underestimates the abilities of the Peking Man.

First, their ability to survive in a cold environment was not poor. As mentioned earlier, the earliest appearance of ancient humans on Dragon Bone Hill was over 700,000 years ago, during a relatively cold period. The cover of the British magazine *Nature* introducing the era of the Peking Man's life explicitly read "Peking Man experienced extreme cold." The fact that the Peking Man could control fire artificially and live in the northern regions of China for an extended period demonstrates their adaptability to the cold.

Second, they were intelligent and bipedal ancient humans with migratory capabilities. Over hundreds of thousands of years, the East Asian *Homo erectus* sometimes settled in Dragon Bone Hill and sometimes left. When the surrounding environment became unsuitable for survival, they would migrate collectively. Even during severe Ice Ages, they could move to warmer environments in the south and then expand northward again after the glaciers receded and the climate warmed.

Therefore, the Ice Age could not have wiped out the Peking Man or the East Asian *Homo erectus* altogether. As a paleoanthropologist questioned, the impact of Ice Ages on climate varies across different regions of the world, and it wouldn't have caused the extinction of East Asian *Homo erectus*. "At least there are no typical glacial relics in southern and central China. Besides, giant pandas, living in those regions, survived; there is no reason why humans couldn't have survived."

If the geneticists' viewpoint is correct, then how were the East Asian *Homo erectus* "exterminated?"

The answer may lie in the "Technological Tragedy" of Tasmania Island.

Tasmania Island is located in the southeast of the Australian continent, separated from the mainland by the more than 200-kilometer-wide Bass Strait. Its area is approximately twice the size of China's Hainan Island. Archaeologists have discovered that during the Ice Age, Tasmania was connected to the Australian mainland by an ice bridge. Human beings inhabited Tasmania around 35,000 years ago. Up until 10,000 years ago, Tasmania was still connected to the mainland, but as the Ice Age ended and the climate warmed, the island became completely isolated from the mainland, and its indigenous inhabitants were cut off from the outside world until European navigators discovered them a few hundred years ago.

When the Europeans first encountered the Indigenous Tasmanians, they noticed that these people did not possess any bone tools such as needles and drills. They also lacked warm clothing, handles for tools, fish hooks, fishing nets, spears, and boomerangs, among other things. The thousands of Indigenous Tasmanians were divided into nine tribes, living a primitive life of hunting and gathering. They used wooden sticks and spears to hunt seals, seabirds, and wallabies for sustenance. However, archaeologists found that the first group of Tasmanian Indigenous people did have bone tools and the knowledge to make them, but they gradually abandoned this technology. Over the course of 10,000 years, their bone tools became increasingly simple, and around 4,000 years ago, they were no longer able to produce any bone tools. Without bone tools, they couldn't sew animal skins into warm clothing. For thousands of years, even in harsh winters, the Indigenous Tasmanians were almost naked, only applying some seal oil on their skin and draping wallaby skins over their shoulders. They used to fish and eat fish abundantly, but this skill was lost around 3,000 years ago, and when the Europeans offered them fish to eat, they found it repulsive.

Despite their lack of certain technologies, the Indigenous Tasmanians did engage in some technological innovation. Around 4,000 years ago, they developed a type of wooden raft made from bundled shrubs. These rafts were either paddled by men or propelled by swimming women. With these rafts, they could travel to nearby smaller islands for hunting. However, the rafts would disintegrate and sink after a few hours in the seawater. Thus, the Indigenous Tasmanians were unable to cross the Bass Strait and make contact with mainland populations.

Over thousands of years, the Indigenous Tasmanians experienced a tragic technological regression, which is referred to as the "Tasmanian technological tragedy." The tragedy resulted from the interplay of population size and technological advancement.

In principle, the larger the population of a group, the more people there are to master various technologies, leading to increased exchange, teaching, innovation, and easier transmission of technology. Conversely, smaller populations are more prone to losing technologies. The early Tasmanian Indigenous population was small and isolated on the island, lacking contact with the outside world. As a result, they were unable to maintain their original technological level, and they helplessly watched their technologies disappear over time.

Similarly, the fate of *Homo erectus* wasn't much different from that of the Tasmanian Indigenous people. During the *Homo erectus* stage, ancient humans engaged in relatively primitive hunting and gathering activities, with group sizes ranging from a few dozen to a few hundred individuals. If there were other *Homo erectus* groups nearby with which they could interact, the network of exchanges could help them maintain their technological level and prevent a decline. However, if nearby *Homo erectus* groups were absent, their technological level was hard to sustain, let alone improve. This is why *Homo erectus*'s stone tool technology progressed very slowly. Over nearly 2 million years of *Homo erectus* history, there were only four or five notable advances in stone tool technology.

The situation of East Asian *Homo erectus* was even more unfavorable. Due to geographical barriers, when *Homo erectus* successfully reached East Asia over 1 million years ago, they unfortunately found themselves trapped in an East Asian version of "Tasmania tragedy." East Asian *Homo erectus* lived in a larger area compared to Tasmania but faced difficulties in exchanging technologies with other regions, leading to isolated development. East Asian *Homo erectus* continued to use the same primitive stone tool technology they brought with them, with some knowledge loss and minor innovations. Compared to the vibrant technological advancements in East Africa and the Middle East, East Asian *Homo erectus*'s technology remained relatively crude.

The technological primitiveness, in turn, limited their group size and their ability to gather and process food. With a weak capacity to sustain larger populations, each group of East Asian *Homo erectus* could only maintain small sizes.

This limited their ability to interact with other *Homo erectus* groups and hindered the exchange and development of technology.

As a result, East Asian *Homo erectus* remained in a prolonged state of survival characterized by primitive stone tool technology and small group sizes. It can be speculated that many of these groups might have perished over time, as even minor events like severe cold, floods, or diseases could easily destroy them. Archaeologists have noted the absence of *Homo erectus* relics in China dating back to 100,000 to 40,000 years ago. This indicates that after a million years of struggling for survival, East Asian *Homo erectus* either became extinct or only a negligible number of groups remained, leaving hardly any trace of their existence.

In the shadow of Tasmania Island's technological tragedy, the ancient *Homo erectus* of East Asia may have long struggled on the precipice of functional extinction. This perturbing state refers to a population so diminished that it can no longer sustain reproduction. A contributing factor lies in their over-reliance on inbreeding, resulting in the inheritance and outbreak of deleterious genes.

Imagine a group of roughly 50 *Homo erectus*, equally divided between men and women, with an average lifespan of less than 30 years. Consequently, only around ten males and ten females would be capable of procreation during the same period. If this isolated *Homo erectus* community lacked genetic exchange with external groups, successive generations would inevitably be closely related, leading to unavoidable inbreeding and a homogenization of genes. This precarious situation could easily culminate in the ruin of the entire community due to the proliferation of disease-causing genes. To avert this functional extinction, maintaining external intermarriage and engaging in widespread genetic exchange with other groups would be essential, mitigating the risks posed by inbreeding.

Even if a handful of *Homo erectus* managed to endure until several millennia ago, they would have faced a sudden influx of a new population—modern *Homo sapiens*—sweeping into East Asia. Confronted with a multitude of modern humans, armed with superior weaponry, the *Homo erectus* population would have been woefully outmatched. They would either have been effortlessly eradicated or succumbed to decline as modern humans claimed their resources.

Confronted with the formidable intellect of *Homo sapiens*, one might wonder if *Homo erectus* had any chance of integrating into their society and passing on their genetic legacy.

Regrettably, such a prospect seems exceedingly improbable.

Over hundreds of thousands of years, East Asian *Homo erectus* and modern *Homo sapiens* (including their ancient African *Homo erectus* ancestors) pursued separate evolutionary paths without any intermingling. This divergence gave rise to what biologists refer to as reproductive isolation. In the presence of only a few *Homo erectus* neighbors, it is conceivable that *Homo sapiens* did not perceive them as fellow beings but rather as a different breed of primates. Likewise, it is not beyond the realm of possibility that *Homo erectus* reciprocated this sentiment and did not regard *Homo sapiens* as their own kind. The enduring period of reproductive isolation led to an insurmountable physical barrier to interbreeding, as well as a profound divergence in behavioral and cognitive patterns, making it unlikely for either group to recognize the other as part of the same lineage.

A past paleoanthropologist once remarked, “Even giant pandas managed to survive; there’s no reason why humans couldn’t.” However, giant pandas’ fortunate survival can be attributed to their not encountering a massive influx of a “new panda tribe” with a much larger population and stronger survival capabilities. Had such a tribe emerged, competing for mountain habitats and viewing pandas as adversaries, their cute existence might not have endured to the present day. Giant pandas were merely luckier than *Homo erectus*.

Turning our attention back to the enigmatic Peking Man of Zhoukoudian, their fate after departing Dragon Bone Hill some 230,000 years ago remains a poignant question. They either met their demise in isolation or valiantly struggled until the time when *Homo sapiens* surged through East Asia, likely culminating in their eventual demise.

The ancestral history of East Asia, now inscribed in the annals of modern *Homo sapiens*, will continue to unfold.

As the Chinese Culture venerates their ancestors, the Peking Man and Dragon Bone Hill hold profound significance for the nation. Their discovery, excavation, and study transpired during tumultuous times of national peril and invasions, providing a beacon of hope and inspiring the populace to rally and defend their heritage. These Peking Man relics emerged as unparalleled “sacred artifacts,” casting a guiding light amidst the darkness and igniting indomitable spiritual strength.

Even before unearthing the Peking Man fossils, intense academic debates about the origin of modern *Homo sapiens* flourished. Over the past century since their discovery, the question of whether the Peking Man is the progenitor of the Chinese people continues to spark spirited discussions within scholarly and public circles.

Regardless of the answer, the profound emotional connection that the Chinese cherish with the Peking Man remains unwavering. For they symbolize an unwavering pursuit of scientific truth, exemplifying the nation’s dedication to rational and scientific exploration, enkindling a profound appreciation for the essence of science and the delight of discovery.

Whether or not the Peking Man constitutes our direct lineage, their legacy remains a cherished treasure within our hearts.

The Column: Do Peking Men Eat or Not Eat People?

In 1943, Franz Weidenreich published his work titled "The New Discovery of Three Skulls of *Sinanthropus pekinensis*," proposing the controversial idea that Peking Man engaged in cannibalism. He suggested that Peking Man hunted and consumed their own kin just as they did with other animals, treating their victims in the same way as they treated animals.

At the Zhoukoudian Peking Man site, a significant number of ancient human skulls were discovered, but comparatively few limb bones were found. Weidenreich argued that natural factors such as water erosion or carnivore predation couldn't explain this phenomenon. Moreover, there was no evidence to support the notion that Peking Man practiced the custom of burying the skulls of deceased relatives. Therefore, the abundance of Peking Man skulls found in the caves might be the spoils of an ancient "headhunting" tribe.

Weidenreich also observed that the Peking Man skull fossils mainly consisted of cranial parts, with very few facial and lower skull bones, a condition that deviated greatly from normal human skulls. Additionally, he discovered some damaged traces on the Peking Man cranial bones and a few limb bones, which he believed were left behind from ancient human brain or marrow consumption. The characteristics of fractured mandibles in Peking Man also suggested possible human-inflicted breakage.

Summing up the various characteristics of the Peking Man fossils, Weidenreich concluded that they belonged to a highly cruel "headhunting" tribe.

His views triggered a long-standing debate, and for decades, numerous paleoanthropologists from both China and abroad presented counterarguments. By 1985, prominent American archaeologist Lewis Binford (1931–2011), refuted the notion of cannibalism among Peking Man based on modern taphonomic analysis and research. Binford believed that the multiple pieces of evidence supporting Peking Man's cannibalism could be reinterpreted using principles of taphonomy.

In essence, over hundreds of thousands of years of burial, Peking Man's remains would inevitably suffer from various natural forces, leading to potential damage and alterations. Binford provided several pieces of evidence to support his perspective:

- The vertical cracks on human limb bones found at the Zhoukoudian Peking Man site were not the result of heavy blows, as Weidenreich claimed, but rather products of bone weathering.
- The two Peking Man bones that Weidenreich identified as evidence of cannibalism, which showed signs of burning, were merely the result of mineral contamination in the soil.

- The peculiar "breakage" in the ancient human mandible fossils from the cave, which Weidenreich attributed to human activity, was a common phenomenon resulting from the weak part of the mandible under pressure or during movement.
- The absence of facial and cranial base bones in all Peking Man skulls, which Weidenreich linked to the consumption of brain tissue, was more likely due to their fragility, making them prone to breakage under various external forces.
- The numerous scratches, incisions, and grooves found on Peking Man skulls, which Weidenreich believed were the product of human action, could also be explained as natural processes occurring during the burial of human remains.

Moreover, Binford discovered bite marks on one Peking Man skull, indicating that carnivorous animals might have been one of the primary collectors of human skulls at the Zhoukoudian Peking Man site. If some Peking Man individuals met an unfortunate fate, they might not have been killed by cruel human actions but rather hunted by ferocious carnivores in the surrounding environment.

CHAPTER 2

Voyage Across Ten Thousand Miles: The Odyssey of *Homo sapiens* to China

The waning sun descended gracefully over the savannas of East Africa, casting a somber hue upon the landscape as the air grew crisper. A nipping breeze rustled, sending a shiver through the chieftain's frame. Even as the tribal leader, his attire of animal pelts mirrored the modesty of his fellow clansmen. Yet, providence smiled upon them this day, as the hunting party returned triumphant with a limping gazelle—a savory repast for the impending night.

Perched upon a rocky ledge, he then embarked on a contemplative journey through the day's hunt. This day was graced with fortune, the bounty ensuring a hearty meal for the tribe. A retrospective gaze painted a picture of changing times; the climate's frigid embrace tightened while the once-abundant prey dwindled. In stark contrast, the tribe's ranks burgeoned. What destiny lay inscribed upon the tribe's scroll of fate?

A distant serenade of ferocious hyenas unfurled on the wind, prompting the chieftain to rise—an epochal resolve was summoned. In the pursuit of the tribe's survival, a momentous decree was etched: a collective exodus from their ancestral grasslands, bound for foreign horizons, a quest for a promised land adorned with sustenance aplenty …

Perchance this vignette unfolded some 70 millennia past, in the heart of Africa—a tribe belonging to *Homo sapiens*, the precursors of modern humanity, in the parlance of molecular biologists. Unbeknownst to the *Homo sapiens* lineage, their imminent odyssey for existence would unfurl an epic cascade of migration across epochs, traversing continents and epochs. Their progeny, traversing terra incognita, would eventually grace every alcove of the world's tapestry, East Asia included.

2.1 Going Out of Africa, One After Another

Before modern *Homo sapiens* migrated out of Africa, there were multiple ancient human species thriving on the Eurasian continent. Before delving into the exodus of modern humans from Africa and their extensive expansion, let's first understand the former inhabitants of the Eurasian continent.

As we learned from the previous chapter, the *Homo georgicus* are the earliest human remains found outside Africa. They lived in the Caucasus mountain region of Eurasia approximately 1.8 million years ago. They might have been among the first *Homo erectus* to leave Africa and one of the first ancient human species to do so. Prior to this, ancient humans were confined to the African continent.

The history of *Homo erectus* spans millions of years, and they might have left Africa multiple times. As early as 1907, archaeologists discovered a fossilized jawbone of an ancient human species in the Heidelberg region of Germany, which was named *Homo heidelbergensis*. Subsequently, traces of this ancient human species were found in many places in Europe and Africa. It is now known that *Homo heidelbergensis* originated in Africa around 700,000 years ago and can be considered a late-stage form of upright-walking human. Some suggest categorizing them separately as an intermediate stage between *Homo erectus* and *Homo sapiens*, although this book continues to include them within the group of *Homo erectus*.

Homo heidelbergensis represents a group of *Homo erectus* that also left Africa. By as early as 500,000 years ago, a group of *Homo heidelbergensis* had already left Africa, entered Western Asia, briefly paused, and then continued westward into Europe. Hence, during this time, there were at least two varieties of *Homo heidelbergensis*—African *Homo heidelbergensis* and European *Homo heidelbergensis*.

The focus of this book is to explore the history of the ancestors of the Chinese people. Therefore, we should pay particular attention to one question: Did a group of *Homo heidelbergensis* venture eastward and reach the territory of China? If so, could they be our ancestors?

Homo heidelbergensis were skilled hunters, potentially one of the first ancient human species capable of hunting large animals. Near *Homo heidelbergensis* sites, archaeologists discovered numerous remains of animals such as deer, wild horses, elephants, hippos, and rhinoceroses, showing signs of butchering. In a *Homo heidelbergensis* site in Germany, a wooden spear was unearthed alongside stone tools and the remains of over ten slaughtered wild horses. In a site in Spain, a finely crafted symmetrical hand axe was found, indicating that *Homo heidelbergensis* possessed advanced stone tool-making skills.

These findings suggest that *Homo heidelbergensis* likely preferred to live in grasslands and forested areas rich with large animals. If *Homo heidelbergensis* chose to settle in Asia, they might have favored the northern regions with vast grasslands and forests that sustained populations of large wild animals—a veritable "meat supply chain" for *Homo heidelbergensis*.

Hence, the absence of *Homo heidelbergensis*'s traces in Central and East Asia by scientists over an extended period becomes more understandable. When considering that *Homo heidelbergensis* also used hand-axe-like tools, and hand axes have been scarcely found within Chinese territory, it seems regrettable to conclude that *Homo heidelbergensis* did not reach China ...

Hold on, let's not jump to conclusions too early.

In 2018, Chinese scientists discovered a skull fossil within the territory of Heilongjiang province, which, based on its features, is likely to be a skull of a *Homo heidelbergensis*. Considering that *Homo heidelbergensis* would have chased prey eastward along the vast grasslands of the Eurasian continent, this discovery might indicate that they also came to the territory of present-day China. However, their activity area was too far north to intersect with other contemporaneous upright humans living in the land of China. Even if a small number of *Homo heidelbergensis* reached the region of Heilongjiang, their descendants would unlikely have settled there and continued to thrive until today.

"Are you suggesting that *Homo heidelbergensis* couldn't be the ancestors of the Chinese?"

Although the Peking Man from Zhoukoudian disappointed us once in our ancestral search, this time, *Homo heidelbergensis* might not let us down. He could indeed be our ancestors' ancestor!

To uncover the connections between *Homo heidelbergensis* and modern humans, molecular biology is our key ally. In the Spanish archaeological site where hand axes were discovered, the leg bone of a *Homo heidelbergensis* was also found. Scientists successfully extracted ancient DNA from the leg bone. Gene decoding revealed that this *Homo heidelbergensis* lived around four to five hundred thousand years ago. This marks the highest record of ancient human age data deciphered through genetics so far.

The genes of *Homo heidelbergensis* also divulge more crucial information. Through the decoding of *Homo heidelbergensis*' mtDNA and whole genome, scientists found that their relationship with two other ancient human species—Neanderthals and Denisovans—is relatively close.

Now, let's get acquainted with these two new "classmates of ancient human."

Neanderthals, also known as the "cavemen," were initially discovered in a cave in the Neander Valley of Germany. The cave entrance was several meters above the ground, and the interior was roughly 5 meters long and 3 meters wide, with a high ceiling. In 1856, quarry workers found some human bones inside and handed them over to local anthropologists, leading to the discovery of

Neanderthals. In the era before genetic research gained prominence, people believed that Neanderthals, based on the comparison of their skeletal features with those of modern humans, were the ancient ancestors of modern humans. They were slightly shorter but robust, living in caves during the Ice Age. Their brain capacity was comparable to or slightly larger than that of modern humans, and their skin was relatively light. These traits were adaptations to life in high-latitude regions. Additionally, they had smooth, long, red, or golden hair. These ancient cave dwellers once inhabited not only the European continent but also regions as far as West Asia and Siberia.

In the 1990s, molecular biologists began examining Neanderthal genes and comparing them to those of modern humans, aiming to uncover their kinship. Due to less advanced technology at that time, only some genes on mtDNA were compared. The results indicated that Neanderthals shared a common ancestor with modern humans around 500,000 years ago. They seemed not to be direct ancestors of modern humans but rather an ancient population mainly distributed in West Asia and Europe. Consequently, scholars at that time believed that Neanderthals were a side branch of human evolution, having no interaction with modern *Homo sapiens*. They were thought to have gone extinct around 30,000 years ago during the process of modern humans' expansion into Europe.

With advancements in genetic technology, our understanding of Neanderthals has changed. In 2010, molecular biologists sequenced a Neanderthal genome and discovered that some of its genes also appear in today's populations of Eurasians. This suggests that interbreeding occurred between modern humans and Neanderthals.

The revelation that modern humans carry Neanderthal genes caused a sensation in the scientific community at the time.

The relationship between modern humans and Neanderthals has become increasingly clear through multiple rounds of testing. Neanderthal genes make up about 1.7% of the modern human genome. Analysis of information extracted from Neanderthal skeletal remains found in Southeastern Europe suggests that interbreeding between modern *Homo sapiens* and Neanderthals occurred approximately 60,000 years ago, around the time when modern humans left Africa and first arrived in Eurasia. In other words, it was when our human ancestors encountered Neanderthals that a significant interaction took place. The reality might not have been as romantic as depicted; it could have been a brutal confrontation: either humans claimed Neanderthals or vice versa.

The interbreeding between modern humans and Neanderthals didn't happen just once. In Southeastern Europe, there's evidence of another interbreeding event occurring at least 40,000 years ago. These multiple instances of interbreeding led to some Neanderthal genes becoming incorporated into the genomes of modern humans and persisting in the genetic makeup of present-day individuals.

Therefore, it's not entirely incorrect to say that Neanderthals are, to some extent, our direct ancestors, accounting for about 1.7% of our genetic heritage.

Denisovans, a later-discovered and more enigmatic ancient human group, add another layer to the story. In the Altai Mountains of Southern Siberia, near the borders of China and Mongolia, there's a cave known as Denisova Cave. In the eighteenth century, a Russian hermit named Denis lived there, giving the cave its name.

In 2008, Russian archaeologists conducted explorations in the cave and discovered a tiny piece of a child's finger bone. This bone fragment was carefully collected and sent to molecular biologists for genetic testing. Due to the cold climate of the Denisova Cave region, DNA preservation was remarkably good. Analysis of the genome from the finger bone revealed that it wasn't from a Neanderthal or a modern human; it belonged to a new human species. Since the bone was found in Denisova Cave, this ancient human was named Denisovan. Because the bone belonged to a young girl, scientists referred to her as "X Woman."

X Woman wasn't the only Denisovan of her kind. In the year 2000, a molar tooth belonging to an adult male was discovered in the same cave, dating back to a similar period as X Woman's existence.

Molecular biologists were particularly interested in understanding the relationship between Denisovans and modern humans. The results of genetic comparisons were once again astonishing. Genetic material from Denisovans was found not only among modern humans originating from Africa but also among people living on Pacific islands, thousands of miles away from the Altai Mountains. Papua New Guineans and Melanesians carried about 4.8% of Denisovan genes in their genomes. This implies that the ancestors of these Pacific island populations interbred not only with modern humans from Africa but also with Denisovans. In essence, modern humans once interbred with Denisovans, leading to the transmission of their genetic legacy through generations and persisting in the genomes of today's individuals. Looking at the entire modern human population, approximately 0.5% of our genes can be traced back to Denisovans.

It appears that Neanderthals were not the sole "paramours" of modern humans.

X Woman was found in Siberia, while Papua New Guineans and Melanesians reside on equatorial Pacific islands, thousands of miles away. From these geographical locations, it's easy to infer that Denisovans and their descendants once had a widespread presence in Asia, ranging from Siberia to Southeast Asia, crossing seas and reaching as far as Indonesia, the Philippines, and other Pacific islands.

The appearances of the two "new classmates," Neanderthals and Denisovans, let us know that while modern humans in Africa had not yet ventured beyond the continent to explore the world, there were already at least three types of ancient humans living outside Africa: various populations of archaic *Homo sapiens*, Neanderthals, and Denisovans. These two new additions to the story might both be descendants of the *Homo heidelbergensis* population that left Africa and reached Asia. Around 400,000 years ago, Neanderthals and Denisovans took different paths in Western Asia, with Neanderthals heading to the European continent and Denisovans venturing further into the Asian interior. Meanwhile, the ancestors of modern humans were still roaming the African grasslands.

Neanderthals were primarily active in the western parts of Eurasia and likely did not reach the territory of present-day China. Denisovans, on the other hand, were active in the eastern parts of Eurasia and certainly lived within the borders of China. Scientists have discovered Denisovan genes within Xizangan individuals, with these unique genes enabling the Zang people to breathe easily in the oxygen-deprived air of the high plateau.

A more significant discovery is that Chinese scientists have found a jawbone fossil believed to be from a Denisovan within China's borders. About 40 years ago, a Buddhist monk entered a karst cave in Xiahe County, Gansu province, and uncovered this jawbone fossil. This cave is located on the Qinghai-Xizang Plateau at an altitude of over 3,300 meters, approximately 2,200 kilometers linear distance from the Denisova Cave in the Altai Mountains. Although ancient DNA wasn't preserved on the fossil, fortunately, degraded ancient proteins were preserved, providing valuable human genetic information. Based on the carbonate deposits on the fossil, scientists estimate that this ancient individual lived around 160,000 years ago. Judging from its genetic information, scientists conclude that this individual was closely related to the Denisovans, suggesting that they might belong to the Denisovan lineage.

If this indeed represents a Denisovan, paleoanthropologists might need to reassess some ancient human remains discovered within China dating back tens of thousands to hundreds of thousands of years. Many of these remains were previously attributed to archaic *Homo sapiens* or sometimes even stretched to be associated with Neanderthals. It's possible that numerous artifacts from that period actually belong to the Denisovans.

The discoveries mentioned earlier confirm that Denisovans are undoubtedly among the ancestors of Chinese people, at least for some portion of the population.

The curtain has been lifted on the exodus from Africa, and some supporting characters have taken their positions on the stage. Now, it's time to officially introduce the protagonists—the African modern humans—as they step onto the scene. Let's explore how our direct ancestors undertook the arduous journey from East Africa to East Asia.

2.2 Flying Eastward Along the Coast

The earliest modern human fossils in the world were discovered in Ethiopia, East Africa, dating back to approximately 195,000 years ago. Due to the difficulty of preserving ancient human fossils, even though these fossils are nearly 200,000 years old, the appearance of modern humans on Earth might have occurred even earlier. Based on the variety and timing of ancient human species in Africa, it is highly likely that modern humans evolved from the group of *Homo heidelbergensis* that remained in Africa. Modern humans, Neanderthals, and Denisovans all share a common ancestry with *Homo heidelbergensis*.

After the birth of modern humans in Africa, there was a span of around 100,000 years during which they remained in their African homeland, honing their skills. The term "*Homo sapiens*" reflects their surpassing intelligence compared to other ancient human species but also primates like chimpanzees. Upon reaching the stage of modern humans, abstract thinking developed rapidly, language capabilities greatly improved, and these advancements brought significant survival advantages.

For instance, their stone tool manufacturing techniques became more intricate, and they excelled at crafting delicate tools from materials such as animal bones, deer antlers, shells, and ivory. They displayed innovation by utilizing novel materials and inventing new technologies. Their diet expanded, ranging from hunting large animals to foraging for wild roots and tubers, making survival in the wilderness effortless. They mastered the use of fire, inventing hearths and distinct living areas. They cultivated rich social cultures and arts.

These wisdom and technologies not only provided survival advantages to modern humans but also led to population growth. Their tribes exceeded the sizes of other ancient human tribes. However, before they could harness collective wisdom and strength, they needed to face the challenges imposed by nature.

Archaeologists have discovered that around 100,000 years ago, a small number of modern humans managed to cross the barriers of the Sahara Desert and ventured northward along the Mediterranean coast of Western Asia. There, they seemed to coexist with local Indigenous Neanderthals for a period. Caves were occupied by both groups intermittently. The exact nature of their relationship remains uncertain. Ultimately, the fledgling modern humans from the Mediterranean coast vanished, perhaps wiped out by Neanderthals or due to natural reasons. This early journey of modern humans out of Africa likely ended in failure.

At that time, Earth was experiencing a harsh period known as the last Ice Age. This Ice Age began around 150,000 years ago and led to significant drops in average temperatures in many regions, about 10 degrees Celsius lower than present temperatures. Subsequent temperature fluctuations persisted, with temperatures becoming colder around 120,000 years ago and lasting for approximately 70,000 years before showing slight improvements.

The Ice Age was catastrophic for ancient human populations on the Eurasian continent, leading to the disappearance of many groups. However, for ancient human populations south of the Sahara Desert in Africa, the tropical location and abundant sunlight maintained relatively higher temperatures. Nevertheless, precipitation was influenced by the Ice Age, resulting in prolonged drought. The tropical rainforests and grasslands in East Africa gradually transformed into deserts and dry grasslands, with only the East African coast benefiting from moist air flows due to oceanic climate.

Consequently, an increasing population of modern human tribes congregated along the more habitable East African coast. This region provided not only familiar forest and grassland foods but also marine resources. In Eritrea, East Africa, archaeologists found abundant debris dating back over 100,000 years, including remains of hunted rhinos and elephants, various shells, and a mixture of human-made stone tools. These findings indicate that modern humans resided here for an extended period, exploiting marine food resources.

Apart from the prolonged Ice Age, certain sudden disasters have also befallen modern humans. Spanning across the vast Indian Ocean from the east coast of Africa, within present-day Indonesia's Sumatra Island to the North, lies Lake Toba, also known as Danau Toba, measuring 100 kilometers in length and 30 kilometers in width. It is, in fact, the caldera of the dormant supervolcano, Mount Toba. Throughout its history, Mount Toba has experienced at least three major eruptions, with the most recent eruption occurring over 70,000 years ago. This eruption was one of the largest volcanic events on Earth in the last two million years, causing copious amounts of volcanic ash and dust to be released into the atmosphere, blocking out the sunlight, plummeting global temperatures, and triggering a massive ecological disaster.

Scientists have evaluated the impact of this catastrophe on the ancient global human population. Neanderthals in Europe were largely decimated by this event, with only a small group persevering in the Caucasus Mountains. The Caucasus region had previously been the home to early hominins who migrated out of Africa, and it once again provided refuge, this time for the Neanderthals.

The Denisovans, who lived in East Asia, also suffered heavy losses, potentially leaving only a few hundred individuals surviving in the Altai Mountains. As for the archaic *Homo erectus*, even though they had not been completely wiped out by the previous lengthy Ice Age, they could not escape the devastation of this cataclysmic event.

Modern humans along the coast of East Africa were also affected by the catastrophe, facing a daunting survival challenge. For migratory modern humans of that era, the adage "where there's bread, there's a homeland" held true. Similarly, for the modern humans congregated along the East African coastline, their homeland was where food could be found.

Crises embody both danger and opportunity. The Ice Age and volcanic eruptions, while bringing about a harsh environment, also led to a considerable drop in global sea levels, estimated to be around 100 meters lower than today. This meant that the Red Sea, which separated the African and Asian continents, became narrower. Additionally, coastlines of continents and islands extended toward the ocean, causing straits between the Persian Gulf, the Gulf of Khambhat (earlier known as Gulf of Cambay) in Southern Asia, and the numerous islands of Southeast Asia to narrow. Many small straits even turned into land bridges.

A "coastal highway" stretching eastward from the East African coastline appeared before modern humans. Following an unsuccessful Western Asian expedition around 100,000 years ago, modern humans embarked on their second journey out of Africa around 70,000 years ago. Could they succeed this time?

Lake Mungo, located in present-day New South Wales, Australia, had dried up around 10,000 years ago. However, at least around 20,000 years ago, the area was lush with water and vegetation. In 1974, Australian archaeologists discovered a male hominin fossil at Lake Mungo, which was later identified as belonging to a modern human. Through repeated analysis, his existence was dated back to around 45,000 years ago. Moreover, beneath the soil layer containing this ancient human fossil, artificial artifacts were found, dating back to approximately 60,000 years ago.

This dating astonished the archaeological community, indicating that the Lake Mungo region had been inhabited by modern humans between 60,000 and 45,000 years ago. This is the oldest modern human fossil found outside of Africa. Archaeologists have found evidence of modern human presence in Southeast Asia's Kalimantan Island and Papua New Guinea, dating back over 40,000 years.

People were immediately puzzled by a question: how could the modern humans found between the east coast of Africa and Australia be younger than the modern humans in Australia?

We know that the Australian continent is surrounded by water, and millions of years ago, it was no longer connected to other continents by land bridges.

The animal populations there are very ancient. Therefore, the ancient humans found on this continent must have migrated from other continents. However, the modern humans discovered by archaeologists in Southeast Asia, the oldest of whom date back only 40,000 years, arrived much later than the analysis of the human fossils from Lake Mungo suggests, which indicates that modern humans reached the Australian continent around 60,000 years ago.

Did modern humans from Africa take an airplane to directly land on the Australian continent? Of course, this is a joke. Archaeologists need a more reasonable explanation. They seek clues about the migration of modern humans by studying the genes of Indigenous people from various regions. From genetic analysis, it is suggested that when modern humans left Africa about 70,000 years ago, they departed from the northeastern coast of Africa. Furthermore, the genes of Australian Indigenous people are complex and ancient, closely resembling those of African populations, which indicates their migration from Africa. These Indigenous people are likely the descendants of the ancient human male from Lake Mungo.

In Southern India, between Africa and the Australian continent, archaeologists have found that the genes of some Indigenous people are very similar to those of Australian Indigenous people. Additionally, the Indigenous people of the Andaman Islands in the Indian Ocean share many characteristics with the San people (Bushmen) and Pygmy people of Africa: short stature, dark skin, and dense curly hair. This is because the Andaman Islands are relatively isolated in the sea and have not been influenced by subsequent waves of other tribes, preserving older genetic traits.

Genetic research also reveals that the modern humans who eventually reached the Australian continent were mainly distributed along the Indian Ocean coast and Australia itself, rather than migrating deep into the continent or along the Pacific coast. Based on genetic information, archaeologists can essentially trace the coastal route taken by modern humans: starting from the northeast coast of Africa, passing through the Indian coastline, traversing Southeast Asian islands, and finally arriving at the Australian continent. Eventually, around 30,000 years ago, they entered the islands where the "Tragedy of Tasmania" from the previous chapter occurred.

Modern humans did not arrive in Australia by plane; they migrated along the "highway" of the glacial-era coastline. Therefore, many modern human sites along the way should be located near the coastline of the glacial period, not the current coastline. After the Ice Age ended, the Earth warmed up, and sea levels rose significantly, submerging many ancient human sites along the coast. This is why archaeologists have been unable to discover these older sites.

So, did the enigmatic "Coastal Rampage" ever grace China's shores via the "Coastal Highway"?

Genetic research has spoken: indeed, they did.

This assembly of "Coastal Rampage" held within them a prowess for navigation, undoubtedly. Australia's continental embrace separated from its brethren millions of years past, rendering its fauna timelessly ancient. Ere the arrival of *Homo sapiens*, the Australian expanse knew not even the presence of primates, let alone antiquated hominids. The likes of Java Man, a form of *Homo erectus*, dwelled on the northern isles a mere 100 kilometers from the Australian coast, yet their gaze yearned for distant lands. Even amidst the ebbing tides of glacial epochs, the journey from Southeast Asia's islands to the Australian mainland entailed a dance with tempestuous waves. Such a feat demanded not only advanced maritime instruments but also a mastery of seafaring crafts that the era lacked.

Their adroit navigational prowess facilitated not only their passage to the Australian continent upon reaching Southeast Asian shores but also their resolute march northward along the coastal tapestry. It is here that China, the realm of today, lay to the north of Southeast Asia. An era, perhaps ancient by our calendar's reckoning—some 50 millennia past—saw the visage of modern *Homo sapiens* emerge along the Chinese coastline. Regrettably, those coastal contours of antiquity now slumber beneath the caress of oceanic depths.

In their wake, these modern wanderers traced the southeastern sinuosity of China's shoreline, traversing ever northward, eventually treading upon the Korean Peninsula and the archipelago of Japan. Among the contemporary denizens of Northeast Asia, among the folds of Mongolia, up to a prodigious half lay claim to the genetic heritage of the Coastal Rampage. This whispers tales of not only a penchant for riding the aqueous crests but also of sojourns deep into continental hinterlands, rendering China's northeastern domains a trodden path for these itinerant *Homo sapiens*.

Perhaps an insatiable yearning for coastal enclaves curtailed endeavors toward inland subsistence. Or could it be that the chill grasp of the Ice Age had rendered the heart of continents too frigid for habitation, thus causing these trailblazing *Homo sapiens* to be scarcely succeeded by posterity? The annals of modern humanity, it appears, are largely inscribed by another cohort of *Homo sapiens*, those who embarked on an exodus out of Africa's embrace.

2.3 Ancestral Hunters of 30,000 Years Past

Approximately 50,000 years ago, a fresh wave of modern *Homo sapiens* from Africa embarked upon a journey. These intrepid souls, following in the distinguished footsteps of their valiant forebears from a century before, once again dared to traverse the formidable expanse of the Sahara Desert. Their destination: the Mediterranean shores of West Asia to the North.

A novel chapter in the annals of human prehistory was thus revealed.

Diverging from the "Coastal Rampage" that set forth two decades prior, these modern humans exhibited an enhanced mastery of hunting. Outshining their ancestral counterparts, the *Homo heidelbergensis*, they earned the moniker "steppe hunters." Archaeological revelations unveil that during the period spanning 60,000 to 50,000 years ago, early humans conceived the world's earliest iterations of contrived projectile tools—nimble spears and bows with accompanying arrows. These instruments of long-range hunting augmented efficacy while diminishing mortality stemming from direct engagements with predatory creatures. The ensuing population surge bore testament to these innovations, precipitating a surge in demographic expansion. It is plausible that this acquisition, especially the adroit manipulation of keen-edged wooden spears and bows with arrows, spurred these pioneering modern humans to forsake coastal routes in favor of pursuing grand quarry toward the realm of West Asia.

Equipped with their nascently conceived bows and arrows, the "steppe hunters" stumbled upon the Eurasian expanse, a true terrestrial paradise. Stretching across vast latitudes, the Eurasian steppe became their "highway of the grasslands," affording swift ingress to the verdant pastures and prolific fauna of Central Asia, still enveloped in a pall of frigidity. Their superlative hunting dexterity facilitated not only survival but flourishing within the chill enclaves of the continent's heartland. In consequence, the indomitable "steppe hunters," establishing their base in Central Asia, embarked upon a ceaseless diffusion in all compass points.

One contingent within this cohort ventured southwards into the Indian subcontinent. Upon reaching the vicinity of the Indian coastline, they chanced upon the "Coastal Rampage" assembly, who had ventured into India more than a decade prior. Even after the passage of such temporal expanse, the "Coastal Rampage" collective remained modest in numbers, unable to mount an effective challenge against the formidable "steppe hunters." Genetic inquiry underscores the integration of female members from the "Coastal Rampage" as constituents of the "steppe hunters" populace, while male heritage was largely terminated.

Could it be postulated that the "steppe hunters" dispatched the males of the "Coastal Rampage," laying claim to their females?

This proposition, while speculative, carries considerable weight. The assimilation processes among disparate ancient societies were unburdened by contemporary human rights and legal constructs. The reigning order adhered to the jungle's dictum, marked by instances of male fatality and female abduction—an evolutionary vestige that continues to echo across human history.

Yet another cadre of the "steppe hunters" ventured eastward, tracing the expanse of the steppe belt. Negotiating the formidable Tian Shan and Altai mountain ranges, the time-honored barriers delineating East and Central Asia were traversed. From a genomic perspective, upwards of 50% of contemporary East Asian populations possess distinctive genetic imprints of the "steppe hunters." Their arrival in Northern China and Mongolia, an event believed to have occurred between 35,000 and 30,000 years prior, left an indelible imprint.

For eons, the vast expanse of China lay dormant and forlorn, its landscape punctuated by a few resilient enclaves of archaic *Homo erectus*. An era of minimal activity and scant migration ensued. Yet, between 50,000 and 30,000 years ago, the landscape metamorphosed into a theater of vibrant engagement. During this epoch, a minimum of two cohorts of modern humans—the "Coastal Rampage" and "steppe hunters"—graced East Asia with their presence. There existed, in all likelihood, vestiges of archaic human communities such as the Denisovans, provided they had not succumbed to the ravages of the Ice Age or the ascendancy of modern humans.

Modern *Homo sapiens* have successfully transcended geographical barriers, repeatedly making their way into the vast expanse of China. This phenomenon finds its roots in their heightened intellect and augmented technological prowess, coupled with the synchrony of opportune celestial occurrences.

Insights gleaned from China's antiquated climatic patterns serve to illuminate this narrative. An analysis of chemical elements within the sediment of the Loess Plateau reveals that between 90,000 and 60,000 years ago, an era of frigid climes prevailed. During this epoch, it is surmised that the entire globe lay ensconced in an icy grip. Notably, it was during this juncture that modern humans, the *Homo sapiens*, astutely seized upon the ebbing sea levels to traverse Africa's shores, instigating an epochal exodus commonly referred to as the "Coastal Rampage." The subsequent span, spanning from approximately 60,000 to 25,000 years past, marked a phase of comparative warmth on the Loess Plateau. This age aligns with the migratory endeavors of "steppe hunters," as they embarked on a sweeping expansion into the Eurasian realm. The period spanning 25,000 to 14,000 years ago is distinguished by an exceptional resurgence of frigidity on the Loess Plateau.

Furthermore, the western precincts of Inner Mongolia once cradled a grandeur expanse of water—Ju Yan Sea (Ju Yan Hai). The lyrical verses of Tang Dynasty wordsmith Wang Wei (699–759) reverberate with the cadence of ages past: "A solitary rider yearns to explore the frontier; Crossing through Ju Yan in vassal dominion. Venturing beyond the Han frontier, homeward-bound geese pierce the barbarian sky." In this eloquent expression, "Ju Yan" evokes the enigma of Ju Yan Sea. The ensuing verse, "The endless desert exudes a solitary wisp of smoke;

The meandering river nestles the setting sun," bears testimony to the existence of Ju Yan Sea during the Tang Dynasty, albeit flanked by an increasingly arid desert panorama. Delving into the vestiges of this ancient lake, scientists discern that during the epochs spanning 37,000 to 34,000 years ago, 31,000 years ago, and 28,000 to 26,000 years ago, the waters surged to elevated heights, emblematic of a temperate and lush climate. Conversely, the phase spanning 19,000 to 14,000 years ago witnessed dwindling water levels, concomitant with a parched atmosphere (potentially compounded by wintry conditions). The salubrious window period around 37,000 to 34,000 years ago marked the epoch when the "steppe hunters" undertook their eastward migration across the Tian Shan and Altai Mountains.

This juxtaposition of ancient climatic shifts with the expansionary trajectories of modern humans alludes to a "teeter-totter effect" within the canvas of East Asian migration. As the global climate assumed glacial mantles, the interiors turned icy and inhospitable to habitation and migration, while the coastal terrains, underscored by receding sea levels, proved amiable, thus engendering the expansive zenith of the "Coastal Rampage" and its lineage in the southern latitudes. Conversely, in eras marked by a planetary embrace of warmth, the interior regions flourished invitingly, even as burgeoning coastal sea levels encroached upon the very fabric of coastal existence, heralding the ascendancy of the "steppe hunters" and their descendants in the northern expanses.

This oscillatory undercurrent unveils the migratory breadcrumbs of modern humans across China's realm through myriad millennia, providing an evocative tapestry to apprehend the ever-changing visage of ancient human existence. Embarking on a sojourn back to Zhoukoudian in Beijing, we stand poised to forge an intimate connection with the primordial ancestors of modern humans, in an attempt to transcend the temporal chasm of tens of millennia.

"Yet, did you not assert that Peking Man of Zhoukoudian is not our forebearer? Did they not depart from there over 200,000 years ago?"

Indeed, the passage of tens of millennia has seen Peking Man vacate Zhoukoudian. Nonetheless, the surge of modern human expansion cast its aegis over Dragon Bone Hill, a "treasure trove" of geomantic significance, rejuvenating its historical vibrancy.

In the annals of 2001, the trowels of archaeologists unearthed vestiges of ancient humans within the embrace of Tianyuan Cave, positioned a mere 6 kilometers distanced from Zhoukoudian's Peking Man site. This repository yielded the antiquities of an epoch, encompassing not only the mandible but also select limb bones, in tandem with a profusion of mammalian remains. Dubbed as the "Tianyuan Man," this embodiment of antiquity, over the course of subsequent years, surrendered its secrets to isotopic scrutiny, elucidating a temporal range spanning from 42,000 to 38,500 years ago. The revelations gleaned from genetic assays spotlighted Tianyuan Man's affiliation with the lineage of modern humans, intertwined with scant traces of Neanderthal and Denisovan genetic heritage.

Viewed through the lens of historical epochs, Tianyuan Man is a probable descendant of the "Coastal Rampage," which migrated northward along the coastline. Despite the present separation of over 170 kilometers between Zhoukoudian and the shores of the Bohai Sea, tens of millennia ago, the coastal boundary of the Bohai Sea stretched from Wuqing to Jinghai in Tianjin. At that time, Zhoukoudian rested merely around 100 kilometers from the coast. From Dragon Bone Hill, extending southeast to the coastal fringe, the landscape lay expansively flat. This dynamic shifted with subsequent marine regression and riverine deposition, effectively propelling the coastline seaward. The journey from the coastal origin to Zhoukoudian likely proved relatively uncomplicated for the descendants of the "Coastal Rampage."

Further indirect substantiation of the link between Tianyuan Man and the "Coastal Rampage" is garnered from a comparison of genetic data with contemporary Amazonian populations of South America. Among the diverse Indigenous groups of the Americas, only the Amazonian people share a genetic closeness with Tianyuan Man. This infers that the forebears of modern Amazonians might share an intimate connection with Tianyuan Man. Given that the migration of Native American populations from Asia is well-established, if only the lineage of modern *Homo sapiens* represented by Tianyuan Man had transmigrated to the Americas, all Native American groups should evince close genetic ties to Tianyuan Man. However, reality diverges as solely the Amazonians, situated in the Southern Hemisphere of the Americas, exhibit a marked genetic proximity to Tianyuan Man. This suggests the existence of an early influx of another modern human group closely related to Tianyuan Man, followed by subsequent migrations. Given the conspicuous maritime capabilities of the "Coastal Rampage," presumed to have extended into the Northeast Asian sphere at an early juncture, it is plausible that the progenitors of Amazonians can be traced back to this tribe, thereby positioning Tianyuan Man as its descendant. The intricate narrative of Native American lineage and expansion is poised to be elaborated upon in forthcoming sections of this discourse.

Furthermore, genetic investigations intimate that Tianyuan Man is unlikely a direct progenitor of contemporary Chinese individuals and potentially left no lasting legacy. The genetic lineage of Tianyuan Man appears to have waned over time, a poignant revelation indeed.

In an effort to console the hearts yearning for ancestral connections, another lineage of antiquated humans emerged atop Dragon Bone Hill subsequent to the epoch of Tianyuan Man. This lineage, famously known as Shandingdong Man, was unearthed in the stalactite caves of the mountain as early as 1933. However, their prompt discovery subjected their original cranial specimens and select bone samples to a fate akin to that of the Zhoukoudian Peking Man, lost amid the turbulence of wars.

Eminently resourceful, Shandingdong Man displayed a penchant for artistry, adorning themselves with a profusion of embellishments, encompassing perforated animal dentition, clam shells, diminutive stone beads, pendants, fish-eye bones, and intricately engraved bone tubes. Their bone implements and ornamentations were imbued with exquisite craftsmanship, testament to their mastery of drilling techniques, capable of perforation from opposing facets.

Shandingdong Man's residential milieu was partitioned into several cavernous chambers. Employing isotopic dating on animal remains within these cavities, researchers ascertained that the upper and lower chambers witnessed habitation approximately 29,000 to 24,000 years ago and 34,000 to 33,000 years ago, respectively. As these animals were likely preyed upon by Shandingdong Man, these temporal ranges align with the era of their habitation.

From the chronicles of Shandingdong Man, their existence coincided with a phase of relative warmth within the Ice Age. Lamentably, the absence of genetic data from Shandingdong Man precludes the molecular biological discernment of their forebears or the survival of their lineage to contemporary times. Nonetheless, their adroit perforation techniques and bone needle implements likely extend from the legacy of the "steppe hunters," who employed bone needles in crafting animal-hide garments. Considering that Shandingdong Man flourished in the wake of the eastward migration of the "steppe hunters," it is highly plausible that they are inheritors of this lineage, distinct from the modern *Homo sapiens* represented by Tianyuan Man, who arrived at Zhoukoudian more than a millennium earlier. Whether these two human cohorts ever crossed paths upon Dragon Bone Hill remains a tantalizing enigma.

"The banners atop the city walls keep changing," the succession of ancient humans upon Dragon Bone Hill serves as a poignant allegory for the ever-shifting tapestry of ancient human settlement across the expanse of China over countless millennia. Guided by the orchestration of climate fluctuations and impelled by the imperative of communal survival, disparate groups traversed the Chinese landscape, ceaselessly striving for the realization of their aspirational tomorrows (refer to Table 2.1).

Table 2.1: Ancient humans from different periods and regions

Stages of ancient human evolution		**African continent**	**Eastern Eurasian continent**	**Western Eurasian continent**
Stage of *Homo sapiens* (400,000 years ago to present)	Late-stage *Homo sapiens*	....	· Shandingdong Man · Tianyuan Man	Cro-Magnons
	Early stage *Homo sapiens*	*Homo sapiens* idaltu	Denisovans	Neanderthals
Stage of *Homo erectus* (Approximately 2 million to 200,000 years ago)		African *Homo heidelbergensis*	· Peking Man · Yuanmou Man · Java Man	· European *Homo heidelbergensis* · *Homo georgicus*
Stage of *Homo habilis* (Approximately 3 million to 2 million years ago)		*Homo habilis*	/	/
Stage of Australopithecus (Approximately 5 million to 2 million years ago)		· *Australopithecus afarensis* · *Australopithecus africanus* · *Robust Australopithecus* · *Australopithecus boisei*	/	/

2.4 From South to North, Travel Across China

Examining the diversity of stone tools owned, we glean insights into the widespread presence of modern *Homo sapiens*, exemplified by Shandingdong Man, across the ancient Chinese landscape of that era. From the northern expanse of North China to the Yellow River Basin, a cohort of these adept individuals engaged in cultural and technological activities akin to those of their prehistoric counterparts. The region we now identify as the Central Plains (Zhongyuan) stood as their vital domain. Divided into distinct tribes, they settled within the verdant valleys of plains, erecting encampments in regions blessed with bountiful water sources and fertile sustenance. These sites evolved into prolonged dwellings, encircled by temporary activity nodes, as these tribes indulged in hunting, foraging, and the artistry of crafting tools from select stones.

Fascinatingly, archaeological endeavors have uncovered vestiges of ancient humanity dating back 30,000 years, south of Zhengzhou. These ancestors transported vividly hued purplish-red quartz sandstone across great distances, meticulously constructing pedestals from them. These platforms held the colossal skulls of straight-tusked elephant (*Palaeoloxodon antiquus*), an immense elephant species whose mature forms weighed beyond 10 tons. The motives behind such endeavors remain shrouded in ambiguity. Whether born of awe, veneration for these colossal beings, or ritualistic entreaties for prosperous hunts and bountiful harvests, the veil of uncertainty prevails. This ceremonial-like spectacle hints at the profound spiritual realm traversed by the ancient humans of that epoch.

The era of Shandingdong Man's zenith extended for millennia, commencing some 25,000 years in the past. It corresponded with the onset of a climatic epoch famously known as the Last Glacial Maximum. Grounded in radiocarbon dating, scholars have calibrated this period between 26,500 and 19,000 years ago. During this Ice Age, approximately a quarter of the Earth's landmass lay cloaked in perennial ice and snow, a stark contrast to the tenth that remains under such cover today. With vast expanses of surface water congealed into frozen terrain, global sea levels retracted by around 100 meters. The temperature panorama reflected a chillier reality, with the average global temperature spanning 5 to 10 degrees Celsius beneath contemporary norms. In latitudes temperate and boreal, the mercury plummeted by as much as 20 degrees Celsius. Equatorial realms, by contrast, experienced a more modest decline of 2 to 3 degrees Celsius, yielding a subdued effect. This extended wintry embrace held firm for 7,500 years, concluding roughly 19,000 years in the past, as the glaciers of the Northern Hemisphere embarked on a gradual retreat, marking the termination of the Last Glacial Maximum.

In the grip of this glacial epoch, the topography of China lay draped in ice and snow. The epoch from 27,000 to 18,000 years ago heralded the emergence of spruce (Picea asperata) and fir (Abies) forests in the Qinling Mountains of Shaanxi province. Flourishing in an environment with mean temperatures ranging from 2.5 to 5.7 degrees Celsius, and situated nearly 3,000 meters above sea level, these arboreal sentinels stood as testament to a bygone era. Akin to their distant counterparts, the present-day climate of the Qinling Mountains hovers around 13 degrees Celsius. Ergo, two millennia past, this landscape bore witness to temperatures 7 to 10 degrees Celsius colder than the contemporary status quo. This shift would have encompassed the northern China territory inhabited by Shandingdong Man, which is estimated to have mirrored a similar 10-degree Celsius drop. The arid chill drove ecological transformations: tall timber and ligneous flora receded while herbaceous varieties burgeoned, especially within the grass family. Concurrently, berry-bearing plants diminished, inflicting considerable hardships upon gathering endeavors.

The advent of the Last Glacial Maximum induced a paradigm shift in the survival strategies of ancient humanity, evident through their choice of stone tools. Blade and microblade tools ascended to prominence, indicative of specialized hunting groups attuned to the frigid, arid landscape, marked by frequent migrations. During this frigid interlude, even if the descendants of Shandingdong Man persisted, it is improbable they continued dwelling in the northern reaches of North China, like Dragon Bone Hill. Plausibly, they embarked on a migratory journey southward. Conceivably, numerous offspring from the Eurasian grasslands, colloquially "steppe hunters," migrated southward to evade the cold, converging upon Chinese soil. These settlers embraced life in the Central Plains or even more southerly domains, lured by milder climates and more abundant sustenance. Much like their historical forebears, the nomadic tribes of the grasslands, these predecessors too ventured southward during icy climatic episodes—albeit their odyssey transpired two millennia earlier. However, bereft of equestrian prowess, their mobility rested solely on pedestrian feats. Their southern migration might not have mirrored the tumultuous upheavals of later times, for roughly 20,000 years ago, Earth's populace remained sparsely scattered.

In summation, compelling evidence suggests a migration narrative wherein ancient humans traversed southward from northern China during the Last Glacial Maximum. This interval signifies a diminution phase for ancestral communities, particularly those hailing from the northern expanses.

So, what spectacle did the southern regions of China present during the era of Shandingdong Man and the ensuing Last Glacial Maximum?

At the juncture when Shandingdong Man and their companions embarked on their expansion in the northern realms of China some 30,000 years ago, the southern expanse of China remained far from tranquil. An early established "Coastal Rampage" lifestyle had taken root in Southeast Asia, with their descendants flourishing in those domains. Simultaneously, the migration of the "steppe hunters" from Central Asia into South Asia was underway, and their lineage gradually permeated Southeast Asia. This amalgamation brought about an assortment of populations in Southeast Asia, or, as articulated by molecular biologists, a panorama of genetic populations inhabiting the area.

Harnessing the temperate climate to their advantage, these groups initiated a northward thrust, infiltrating the southern regions of China. Their ingress could be traced along two trajectories: one traversing a "coastal highway," making inroads from the southern sphere deep into the heartland, and the other potentially meandering through the hinterlands of Southeast Asia, tracing river courses northward to eventually reach the present-day Yunnan region. Molecular biologists unearthed ancient genetic strains persisting in modern populations in Myanmar, concurrently surfacing in contemporary communities of Southwestern China, thereby attesting to a linkage between these vicinities. Through genetic dissection, scholars approximated that ancestral groups from Myanmar migrated into Chinese territory roughly 25,000 years ago, if not more recently.

Molecular biologists, employing a holistic approach, have retraced the expansive journey of modern *Homo sapiens* across the Chinese mainland, amalgamating data from Y chromosomes and mtDNA of contemporary populations to chart both paternal and maternal genetic lineages. Their scrutiny unveiled an alignment between the distribution of particular Y-chromosome genotypes and distinct mtDNA genotypes, hereby denoted as "Cluster 1" and "Cluster 2." The former predominated among modern populations in the north, while the latter held prominence in their southern counterparts. Moreover, given the near-equivalent genetic diversity inherent to both clusters, it was surmised that ancient populations harboring these clusters traversed into China's dominion around identical epochs.

In a departure from expectation, populations embodying these clusters did not progress into China dichotomously from polar directions. Molecular biologists discerned that both originated from the southern sphere and subsequently advanced northward in a graduated manner. However, Cluster 2's southward expansion had commenced earlier, accounting for its prevalence among a multitude of contemporary populations in the South. Meanwhile, Cluster 1's foray into Chinese territory was followed by a pronounced surge after their migration to the northern precincts, culminating in its prevalence among various modern populations in the north.

The genetic lineage genuinely stemming from the Mongolian Plateau and Siberia—emanating from the ancient populations journeying southward from the Eurasian steppe—constitutes a minute proportion within the spectrum of today's modern populations. The destiny of these ancient northern populations, compatriots of Shandingdong Man, seemingly harbored less optimistic prospects.

Let us indulge in speculation regarding the migratory course of these ancient populations on the Chinese mainland following the cessation of the Last Glacial Maximum approximately 19,000 years ago.

Under the aegis of the Last Glacial Maximum's impact, the ancient populations in the North bore grievous losses. This entailed not only a substantial decrement in their populace and tribal conglomerates but also the forfeiture of numerous unique genetic strains. Analogous to the annals of multiple significant extinction episodes in Earth's annals, the Last Glacial Maximum bequeathed the extinction of myriad ancient populations. In stark contrast, the southern ancient populations sustained less adversity during this epoch, with their populace, tribes, and genetic diversity surpassing their northern counterparts. As the Last Glacial Maximum waned and a relatively temperate interlude ensued, the flourishing southern ancient populations embarked on a "northward expansion," charting their course toward the reinvigorating Central Plains and venturing even farther North.

In today's China, among various ethnic groups, ranked by genetic diversity from highest to lowest, the order is as follows: Southern non-Han ethnic groups (i.e., various minority groups in the South), Southern Han ethnic group, Northern Han ethnic group, and Northern non-Han ethnic groups (i.e., various minority groups in the North). Despite the ongoing ethnic integration in Chinese history, there still exists a genetic North-South distribution difference, which is a trace left by the ancient Southern population's expansion toward the North. Where genetic diversity is richer, that is the homeland, the starting point of migration.

In a parallel thread, the distribution of dental patterns among disparate populations offers insights into the trajectories of ancient population expansion. Well before molecular biologists embarked upon the modern study of human migratory routes, archaeologists had already embarked on an expedition of exploration through the distribution of dental patterns.

Archaeologists have found that dental patterns of ancient populations in Eastern Asia can be classified into two types: "Sinodonty" and "Sundadonty." The dental patterns of populations in Northern China, Mongolia, Korea, Siberia, and even Native Americans mostly belong to the Sinodonty type, while populations in Southeast Asia and the Pacific islands have dental patterns belonging to the Sundadonty type. In Japan, the majority of people have Sinodonty type, while the dental patterns of ancient Jomon people and modern Ainu people belong to the Sundadonty type. Populations in South China seem to be situated in the intermediate zone between the two dental pattern types.

Based on the distribution of dental patterns, archaeologists propose that the Sundadonty dental pattern might have emerged in Southeast Asia around 30,000 to 17,000 years ago, and then this population with such dental patterns advanced northward along the East Asian continent. Due to the Ice Age conditions at that time, coastlines extended toward the continental shelf direction, connecting Japan, China's Taiwan island, China mainland, and many Southeast Asian islands to the mainland. Another branch of ancient Southeast Asian population later moved northward from the inland areas of the continent, generating the Sinodonty dental pattern in Northern China, possibly around 25,000 to 18,000 years ago, and then they continued northward into Siberia and even crossed the sea to the American continent.

The correlation between archaeological delineations of dental pattern dispersion and molecular biologists' genetic inferences concerning the southern population's northern diaspora is evident. The Sundadonty cohort seems to mirror the essence of Cluster 2, while the Sinodonty cohort may find resonance within Cluster 1.

Furthermore, an observation of note is the apparent limited genetic interchange between Cluster 1 and Cluster 2 over an extensive time horizon—a testament to their preservation of distinctive genetic identities. This intriguing phenomenon enhances our comprehension of the migration patterns of our modern human forebears, a tapestry woven across the canvas of millennia.

In comparison to their predecessors, *Homo erectus*, modern *Homo sapiens* shine forth with heightened intellect and refined technological prowess. This evolution has bestowed upon them superior adaptability, resulting in increased survival capabilities. Their numbers have burgeoned, with larger collectives comprising each group, and their innate migratory prowess has been honed to a remarkable extent.

In essence, the ascent of human evolution to the echelons of modern *Homo sapiens* has largely emancipated them from the specter of the "Tragedy of Tasmania" and the looming threat of functional extinction, except for a few ill-fated enclaves that encountered the aforementioned technological calamity.

Geographical impediments, once perceived as impassable divides segregating East Asia from other regions of Eurasia, have been relegated to insignificance in the face of the inexorable march of modern humanity.

Their odyssey of migration unfolded against the backdrop of sprawling, pristine landscapes. The chronological expansion of ancient human civilizations during this epoch can aptly be likened to a systematic "filling-in-the-blanks" paradigm. Nature herself had furnished numerous enclaves amenable to the aspirations of modern *Homo sapiens*, thus providing an opportune canvas upon which they multiplied and ventured over several millennia, meticulously populating these untamed spaces.

Sequestered in their respective abodes, ancient populations embarked on divergent evolutionary trajectories, unfurled across the canvas of time. The vast tapestry of Earth's expanse, coupled with infrequent genetic intercourse between scattered enclaves, nurtured a pronounced autonomy in evolution, bequeathing unique genetic footprints to each populace. Termed "genetic drift" within the annals of molecular biology, this phenomenon finds its semblance in the diaspora of a larger cohort into smaller fraternities. Over the march of time, these microcosmic assemblies, germinating from the original, inherited genetic stock, transformed into an array distinct from the larger corpus, concomitant with a variegation in the genetic tapestry between the emergent subgroups.

The trajectory of modern *Homo sapiens*, post their exodus from Africa, charts the history of genetic evolution through ages. A comprehensive diaspora originating from a larger assemblage continually segues into intimate clusters, each staking claim to discrete pockets of survival, catalyzing autonomous evolution within their newfound sanctuary. During this phase, the vector of genetic drift and fragmentation surpasses the trajectory of genetic exchange and amalgamation. The result culminates in the proliferation of diverse genetic compositions strewn across the expansive theater of China's dominion—a celestial constellation of ancient lineages.

Commencing around 70,000 years ago, Earth's chronicles heralded the arrival of its most recent glacial epoch. This era culminated approximately 20,000 years ago in the acme of frigid dominion. Subsequently, the planet commenced a gradual abatement from the icy grip. As the Earth's embers rekindled, lands exhaled, particularly within the recesses of the heartland. The advent of spring unfurled verdant landscapes along riverbanks, as temperate breezes intoxicated ancestral spirits. Harnessing this ethereal warmth, our forebears—modern *Homo sapiens*——embarked on a grand odyssey, a relentless northward and southward traverse. Their footfalls echoed across the vast canvas of the Chinese mainland, establishing abodes from the polar extents to the temperate embrace. These modern ancestors, distinguished by sagacity and capability, emancipated from nature's yoke that had previously ensnared their *Homo erectus* antecedents. In the East Asian theater, modern *Homo sapiens* are galvanized, evolving into an embodiment of robust potency, imbibing novel skills, and orchestrating their unique destiny.

The Column: The Genetic Bequest from Ancient Cave Dwellers

Approximately 1.7% of the genes in modern humans come from Neanderthals. In other words, our *Homo sapiens* ancestors interbred with Neanderthals. In fact, it seems that *Homo sapiens* were quite fond of "mixing and mingling," as they also interbred with Denisovans. Genes from Denisovans are found in various populations, including the Zang people, Yi people, Sherpa, Melanesians, and Indigenous people of Papua New Guinea. On average across modern humans, about 0.5% of our genes come from Denisovans. The genes that allow highland populations to adapt to low-oxygen environments, such as those found in Zang people and Sherpas, are attributed to Denisovans.

Neanderthals, like Denisovans, were also quite "romantically adventurous." After branching off from their potential common ancestor, *Homo heidelbergensis*, they interbred with each other. At least 0.5% of Denisovan genes can be traced back to Neanderthals. Denisovans also interbred with another unknown human species. This unidentified species might represent another branch of *Homo heidelbergensis*.

Among the genes inherited from Neanderthals is one associated with the synthesis of keratin, a fibrous protein present in our hair, skin cuticles, and nails. It is believed that when early modern *Homo sapiens* migrated out of Africa, their hair was likely curly. Some Eurasian populations today possess straight hair, attributed to the presence of the Neanderthal gene.

Furthermore, a gene involved in skin pigment synthesis also comes from Neanderthals, contributing to lighter skin color in Eurasian populations, setting them apart from their African counterparts. Lighter skin color provides a significant survival advantage in higher latitudes where sunlight is less abundant. Lighter skin is more efficient at producing vitamin D under conditions of limited sunlight, reducing the risk of diseased conditions like Rickets and Osteomalacia.

The ancient choreography of interspecies couplings remains a delicately rendered artistry, more mosaic than mass occurrence. The tapestry of interwoven lineages bears the traces of a few ephemeral encounters between different human species. Nonetheless, those fortunate individuals who cradled the genetic blessings of their ancient ancestors found themselves endowed with a unique advantage, ensuring that these blessings radiated across generations. Thus, modern denizens of the Eurasian expanse stand as living testaments to the subtle echoes of Neanderthal heritage etched into their genetic narratives.

Yet, within this legacy, not all is a tale of benefaction. Some of the genes passed down from Neanderthals carry with them the seeds of vulnerability, rendering us more susceptible to certain afflictions—type 2 Diabetes, Lupus, and the lingering specter of Crohn's disease, a chronic ailment that besets the bowels. Furthermore, scientists speculate that specific Neanderthal genes may contribute to fertility disorders. Consequently, there are minimal Neanderthal genes present on the X chromosome of modern humans.

CHAPTER 3

From Northern Millet to Southern Rice: The Harmonious Serenade of Huaxia Farmers

Emanating from the heart of the Middle Kingdom, the earliest symphony to grace the airwaves may well have been birthed by the delicate trill of a bone flute.

In Jiahu, Henan province, at the site of Jiahu Lake, dozens of bone flutes made from the wing bones of red-crowned cranes, dating back to around 9,000 to 7,800 years ago, were unearthed. These bone flutes represent the oldest musical instruments ever discovered in China.

The Jiahu bone flutes, elegant in their slender dimensions—spanning just over 20 centimeters with a mere 1.1-centimeter diameter—boast symmetrical perforations along a single curvature, seven in all. Some hold the enigmatic addition of minuscule apertures adjacent to primary tonal apertures, conceivably tuning orifices orchestrated by the ancients. Enchanting to ponder, this bespoke an intricate craftsmanship where harmonious balance was refined, honed through painstaking trial and error, manifesting a mastery of auditory artistry.

Amid epochs when material scarcity dominated the realm, one might wonder whence the luxury of musical dalliance emerged.

The response lies within their alchemical prowess: the alchemy that transmuted wild flora into delectable repasts, and feral fauna into the gentle companions of hearth and home. Pioneers of "agrarian cultivation," these visionary progenitors embarked upon China's agricultural renaissance, etching an indelible chronicle across the vast expanse of China.

3.1 Foxtail and Proso Millet: The Foundation of Northern Ancestors' Civilization

"Abundant granaries lead to understanding of etiquette."[1] Only when one's stomach is full, people pay attention to social etiquette, culture, and the arts—Jiahu bone flutes were crafted and played by people.

Before the advent of agriculture, ancient humans obtained food through fishing, hunting, and gathering. Farming was a labor-intensive way of life, especially considering the low initial crop yields at the beginning of agriculture's emergence. Working in the fields was even more arduous. The protein content of food obtained from farming was lower and less flavorful compared to game and fruits. The only advantage farming had was its ability to sustain larger populations. Under the same amount of land and labor time, agricultural activities provided ancient humans with more food, or more energy, to feed more people, even though the quality of the food was somewhat inferior.

Comparing farming with other modes of production gives us insights into the reasons behind the emergence of agriculture. The ancient choice of the more labor-intensive farming was likely out of necessity, possibly due to environmental pressures.

Therefore, before tracing back to our first batch of agricultural ancestors, we need to rewind a bit and understand the global climate changes following the Last Glacial Maximum that marked the end of the last Ice Age.

Around 18,000 years ago, during the Last Glacial Maximum, was the coldest period. Over the subsequent 4,000–5,000 years, the Earth gradually warmed, and by around 14,000 years ago, a period of substantial warmth emerged, causing a rapid rise in sea levels. Then, between about 12,800 to 11,500 years ago, a cooling event known as the "Younger Dryas" occurred, lowering global temperatures by around 7 to 8 degrees Celsius. Afterward, the Earth warmed again, until around 8,400 years ago when another cooling event occurred, although it was milder than the Younger Dryas.

The dawn of agriculture on the Chinese mainland dates back to the Younger Dryas Period and the subsequent warming period. The impact of the cooling event was much greater on ancient people in Northern China than on those in the south. Therefore, let's first understand the "self-rescue through production" efforts undertaken by northern ancient people in response to the cooling event.

During the warm period that started around 14,000 years ago, the ranges of many large animals living in cold climate zones moved northwards. The primary

---

[1] 仓廪实而知礼节. from Biographies of Merchants, *Shiji* by Sima Qian《史记·货殖列传序 —Translator's note.

hunting targets for ancient people in the north became warm-weather animals like wild boars and deer. The abundance of food in the warm climate supported a larger population. However, the Younger Dryas Period suddenly arrived, causing harsher natural conditions. Many large animals saw reduced numbers or went extinct, drastically reducing the number of available prey for hunting. To avoid starvation, the ancestors turned their attention to the vast fields of wild grass.

Green foxtail (*Setaria viridis*) and hairy cupgrass (*Echinochloa crus-galli*) are annual grasses of the Poaceae family that widely grew in Northern China, producing tiny grains. During warmer times when animals and berry-bearing plants thrived, the northern people might have overlooked the seeds of these two grasses. But in colder times, they had to cherish anything edible in nature, even the small seeds of green foxtail and hairy cupgrass. According to agronomists, their diet at this time was "broad-spectrum," meaning they ate whatever was available.

As clever modern humans, they not only gathered a wide range of plant seeds but also considered selecting seeds from plants with higher yields and plump grains. They scattered these seeds on the land, allowing them to grow freely, hoping for greater yields in the future. In this way, they took the first step from gathering to farming.

Green foxtail is the wild ancestor of foxtail millet (粟, *Setaria italica*), commonly referred as "xiǎo mǐ" (小米, "little rice" in modern Chinese), which is used to make porridge in our daily lives. On the other hand, hairy cupgrass is the wild ancestor of proso millet (黍, *Panicum miliaceum*), known today as "huáng mǐ" (黄米, "yellow rice" in modern Chinese). While proso millet is rarely cultivated in modern agriculture, it still has its place in certain Northern Chinese dishes like "Lǘdagunr" (驴打滚, glutinous rice rolls with sweet bean flour) and "niandoubao" (黏豆包, steamed bun stuffed with sweetened bean paste).

Both foxtail millet and proso millet are crops native to Northern China, but scholars haven't fully determined the exact locations and times of their earliest domestication. Initially, it was believed that the Yellow River Basin, with its favorable natural conditions, especially in the plains suitable for irrigation, might have been the origin of these millets. However, as both foxtail and proso millets were domesticated from wild grasses, their initial areas of improvement might not necessarily have been suitable for large-scale farming; instead, they might have been areas where these wild grasses thrived, like some mountain valleys. Additionally, the superior natural conditions in the Yellow River Basin might have hindered the initiation of primitive agriculture, as the people there might not have faced significant survival pressures to modify these "wild grasses." In regions with less favorable conditions, such as the colder climates of the Yanshan Mountains or the Liao River Basin in the northeastern areas, people might have been more motivated to domesticate these wild grasses to supplement their food sources.

Plant archaeology is starting to provide evidence for these hypotheses. For example, botanists have found charred foxtail millet grains that already show signs of cultivation in the Donghulin site near Beijing. However, these grains were still very small in size, suggesting that people of that time were in the transitional phase of domesticating green foxtail to foxtail millet. Similarly, other botanists have discovered domestication-stage seeds of proso millet in the Shizitan site in Southwest Shanxi, while foxtail millet was still in its wild state. In contrast, the domestication of foxtail millet in the Yellow River Basin likely occurred around 8,000 years ago, which is later compared to more northern regions.

Interestingly, the domestication origins of foxtail and proso millets might also have been different. Considering their characteristics, foxtail millet requires more moisture and tends to grow in relatively wetter areas, whereas proso millet thrives in drier environments. Plant archaeology findings suggest that the cultivation of proso millet first appeared in the transition zone between the forest and grassland vegetation in Northern China. It's most likely to have originated in the western parts of Liaoning, the Xar Moron River Basin in Inner Mongolia, and the northern foothills of the Yanshan Mountains—areas with drier climates and precipitation levels around 400 millimeters. The earliest cultivation of foxtail millet likely took place in slightly southern regions, potentially in the northern parts of the Taihang Mountains to the southern parts of the Yanshan Mountains. These areas experience wetter conditions, with precipitation levels also around 400 millimeters.

Let's glimpse into their crop domestication process through an ancient "granary."

In the southern part of Hebei province, on the eastern foothills of the Taihang Mountains, there's a site called Cishan. Through multiple archaeological excavations, hundreds of pit dwellings have been discovered, nearly 200 of which were used for storing grain. In the first excavation alone, 88 grain pits were found, storing approximately 109 cubic meters of grain, equivalent to around 69,000 kilograms. When including the grain pits discovered in later excavations, this ancient grain base boasted an astonishing storage capacity.

Through the analysis of the grain in the granary, we have determined that these grains were harvested and stored between 10,000 and 8,700 years ago. So, what kind of grains were stored in these ancient granaries?

The grains from the Cishi Site have long since decayed. When initially unearthed, they were a faint gray-green color, which turned gray-white after air-drying. Nonetheless, the distinct outer shells of the grains are still discernible. Although the seeds themselves have decomposed, plant scientists know that when plants are alive, they absorb trace amounts of soluble silica while taking in water. This silica accumulates throughout various parts of the plant, and after the plant dies, it forms corrosion-resistant phytoliths. These phytoliths retain the plant's morphological characteristics, enabling researchers to deduce the plant's species. While phytoliths are not the plant's genes, they store similar information about the plant.

By studying the phytoliths in the grains from the Cishi Site, plant scientists discovered that in samples dating back to 10,300 to 8,700 years ago, the phytoliths of charred grains indicated that the granaries contained entirely proso millet. From 8,700 to 7,500 years ago, there was a small presence of foxtail millet phytoliths appearing in the granaries. However, the content of foxtail millet never exceeded 3% of the total, and proso millet remained the primary grain. This implies that the charred crop samples from the Cishi Site were primarily composed of proso millet. In ancient times, people in the northern regions consumed proso millet before domesticating foxtail millet and incorporating it into their diet.

Why was the granary initially filled with proso millet instead of foxtail millet?

Firstly, this can be attributed to proso millet's hardiness compared to foxtail millet. Proso millet has a short growth cycle and is more drought-resistant and tolerant of poor soil conditions than foxtail millet. The ancient people predominantly lived in valleys with water sources, as those areas were suitable for fishing, hunting, and gathering. Given their limited technological capabilities and the absence of large expanses of arable land, proso millet became their preferred supplementary food source. Additionally, climate change likely played a role in proso millet's early prominence. As mentioned earlier, the onset of colder periods necessitated the rapid identification of new food sources. The plant that was easiest to domesticate would fill the granaries first, and proso millet was the choice due to its ease of domestication and cold tolerance.

The domestication of crops, from wild varieties to cultivated forms suitable for farming, is a gradual and evolving process. Throughout this process, ancient people improved individual plant grains through generations of selection, resulting in larger grain sizes and higher yields per unit area. If we estimate that proso millet was essentially domesticated around 10,000 years ago, the basic domestication of foxtail millet likely occurred around 8,000 years ago. Many ancient sites from that period, such as those in the Liao River Basin, the upper and middle reaches of the Yellow River, exhibited characteristics of domesticated foxtail millet.

Although the domestication of foxtail millet posed more challenges than proso millet, once successful, it provided a higher yield. If water and nutrient conditions were sufficient, foxtail millet outperformed proso millet in terms of yield. Over time, foxtail millet gradually replaced proso millet and became the primary staple in ancient people's granaries.

Initially, foxtail millet was also cultivated in valley areas. Some ancient Chinese legends even offer clues about its cultivation. For instance, in ancient texts mentioning the "Chinese Agricultural God"—Shennong (神农氏)—it is said that "the Shennong City is located at Yangtou Mountain, and there is the Shennong Spring below the mountain, where Shennong obtained the place of fine grains."[2] These grains primarily referred to foxtail millet, and the legend describes Shennong planting foxtail millet on hillsides. This likely provides an early depiction of ancient people cultivating foxtail millet. In the pre-Qin Period, the term "Shè Jì" (社稷) was often used to symbolize the entire world, where "Shè" (社) represented the deity of land, symbolizing the nation's territory, and "Jì" (稷) represented high-quality grains—foxtail millet, for example—which formed the fundamental food source for the people of that time.

After cultivating foxtail millet, the agricultural significance of proso millet diminished. Nevertheless, it still maintained a secondary role in the granaries of northern regions, serving as an important supplementary food source. One significant contribution of proso millet that continues to influence us is its establishment of China's ancient and contemporary length unit—the "chǐ" (尺). After the unification of China by the First Emperor of Qin (221–206 BCE)—Qin Shi Huang, standardizing measurements became imperative. Therefore, a medium-sized proso millet grain was chosen, and a standard length was defined as the span of 100 grains placed horizontally. This length unit, known as "chǐ," was then used to create measuring vessels for grains, subsequently unifying mass units. Additionally, the chǐ length unit was employed to assist in the formulation of the lunar calendar.

As you can see, we still inhabit a world defined by the measurements set by proso millet.

Another important crop originating from Northern China is the bean (菽, shū), commonly known as the soybean today. Compared to grain crops like foxtail millet and proso millet, beans excel in their high protein, fat, vitamin, and mineral content. The protein content of soybeans is six to seven times that of the other two grains of the same weight. Similar to other leguminous plants, soybeans develop nodules on their main and lateral roots that host nitrogen-fixing bacteria, allowing them to synthesize protein using atmospheric nitrogen. Consequently, soybeans held exceptional significance for ancient populations that lacked animal protein sources, as they could obtain a substantial amount of plant protein from soybeans for nutritional supplementation.

---

2 "长子县羊头山下有神农泉，北有谷关，即神农得嘉谷处。"《魏书·地形志 —Translator's note.

The domestication of beans might have been later than that of foxtail millet, with the exact time and location of domestication still unclear. Wild soybeans were widely distributed from the Greater Khingan Range in the northeast to the Yunnan-Guizhou Plateau in the southwest. In oracle bone inscriptions, occurrences of "the harvest year of bean" and "the harvest year of proso millet" are divined simultaneously, indicating that the planting and harvesting seasons of these two crops were similar in ancient times. This suggests that beans might have been domesticated in regions close to where proso millet was domesticated. Therefore, scholars speculate that areas with good sunlight such as mountainous and basin regions near the Yanshan Mountains in the northern part of China might be the origin of soybean cultivation.

Around 4,000 years ago, cultivated soybeans derived from wild soybeans appeared in the Liao River Basin, the Yellow River Basin in Northern China, and even in the Yangtze River Basin in the south. Furthermore, primitive oil extraction techniques emerged, indicating that ancient people recognized the nutritional value of soybeans.

Foxtail millet, proso millet, and beans all originated from Northern China, and their linguistic pronunciations are quite similar.[3] This leads to speculation that these pronunciations might have been a unified term used by ancient populations for edible plants or plant seeds in ancient times. As these plants were successively domesticated, their pronunciations gradually diverged.

During the Neoglacial and subsequent warm periods, there were similar climate changes across the Eurasian continent. After foxtail millet, proso millet, and beans were domesticated in Northern China to provide sustenance and nutrition, these crops rapidly spread from China to other regions.

Around 7,000 years ago, British archaeologists discovered traces of proso millet in more than 20 different locations across Europe, spanning from the western coast of the Black Sea to Eastern and Central Europe. Undoubtedly, this was the result of the westward spread of domesticated proso millet from China.

The dissemination of foxtail millet closely followed that of proso millet. The earliest physical evidence of foxtail millet in Europe dates back to around 5,000 years ago, and there was a noticeable increase in its cultivation around 3,000 years ago. Foxtail millet likely spread into Europe through the Eurasian steppe, entering through the coast of the Black Sea. Foxtail millet also spread eastward to the Korean Peninsula and the Japanese archipelago. For instance, foxtail millet was excavated from a site in South Korea dating back around 4,500

3 The Chinese pronunciation (*pinyin*) of Foxtail millet "sù"; the pinyin of proso millet is "shǔ"; the pinyin of bean is "shū."

years, and it was found in archaeological sites in Hokkaido, Japan, dating back to around 4,000 years ago. Foxtail millet even spread southwest to the distant ancient India. Linguists have found that the Sanskrit word "Cinake" for foxtail millet was also used to refer to China. Foxtail millet likely spread to ancient India through the river valleys of China's Southwestern Hengduan Mountains, passing through the southern corridor of the Himalayas.

Therefore, foxtail millet and proso millet not only filled the stomachs of ancient people in Northern China but also fed the people across the vast Eurasian continent and surrounding islands. Since food is the foundation of life and culture, and a thriving civilization requires abundant food, can the westward spread of China's foxtail millet and proso millet around 5,000 years ago be considered a significant impetus from East Asia to Western civilization?

3.2 From *Oryza Rufipogon* to *Oryza Sativa*: The First Scent of Rice in Asia

When it comes to the contributions of East Asia to world civilization, how could we overlook the crystallization of wisdom from the ancient people of the southern regions—rice (*Oryza rufipogon*)?

There used to be diverse opinions about the origin of rice. The earliest Western scholars to address this question found that the term "dào" (稻), meaning rice, in various Western languages originated from Sanskrit in India. Coincidentally, India was an area where wild rice (*Oryza rufipogon*) was abundant. Therefore, they speculated that India might be the birthplace of rice. Later, a significant amount of wild rice was discovered in Southeast Asia and within China, which redirected scholars' attention to these areas. In the 1950s, scholars proposed that the initial domestication of rice might have occurred in Southeast Asia, given its diverse flora and fauna, which provided ample resources for selecting and hybridization experiments.

All of these viewpoints were challenged. As discussed earlier, regions with superior natural environments and lower survival pressures would not be the prime locations for the emergence of crops. Southern and Southeast Asia were rich in animal and plant resources, and the ancient people engaged in fishing, hunting, and gathering rather than the labor-intensive agriculture. It's only after crops had been domesticated over an extended period, and yields had become sufficiently high, that they spread to these regions and motivated their cultivation.

Whether through theoretical discussion or practical archaeological excavations, scholars eventually turned their focus to the Yangtze River Basin in China while exploring the origin of rice. In the 1990s, archaeologists discovered artifacts of cultivated rice dating back approximately 8,000 years near the Chengtoushan site in Northern Hunan, along with remnants of rice paddies, including primitive irrigation systems with water holes and channels. This possibly represents the earliest evidence of well-irrigated rice fields.

Subsequent archaeological findings were even more exciting. At sites such as Xianrendong and Diaobitanhuang in Northern Jiangxi, evidence of wild rice phytolith dating back 12,000 years and evidence of cultivated rice phytolith dating back 10,000 years were discovered, indicating that ancient people in that region were engaged in rice domestication activities over 10,000 years ago. In 2004, evidence of domesticated rice dating back 18,000 to 14,000 years was found at the Yuchanyan site in Hunan, making it the earliest known rice remains in the world.

Consequently, contemporary scholars have largely accepted that Southern China is the birthplace of rice.

Cultivated rice is mainly divided into two major types: Indica (籼稻, xiāndào) and Japonica (粳稻, jīngdào). Indica rice is suited for growth in the humid and warm regions of the south. The Yangtze River Basin and areas to the south mainly cultivate Indica rice. Japonica rice, on the other hand, thrives in the dry and cool regions of the North. The Yellow River Basin and the northeastern regions of China are the primary areas for cultivating Japonica rice. Botanists, through genetic analysis of rice, believe that Japonica rice appeared first. For example, ancient rice dating back 7,000 years was discovered at the Hemudu site in Zhejiang. Comparative genetic analysis showed its similarity to the majority of Japonica rice genes, suggesting that the ancient people of our southern regions initially consumed Japonica rice, similar to the primitive variety of northeastern rice today. It is likely that the initial domestication of Indica rice occurred later in the ancient plains of the Ganges in India. Evidence of rice consumption by ancient people was found in sites dating back 7,000 to 5,000 years. Subsequently, Chinese Japonica rice varieties spread to ancient India, hybridized with the original Indica rice, and improved the latter, completing the domestication process of Indica rice.

Similar to the protracted domestication processes of proso and foxtail millet in the northern territories, the intricate evolution of rice in the southern regions also spans millennia. In the contemporary landscape, the Jiangsu and Zhejiang provinces stand as cradles of early ancient civilizations in China, where the nuanced taming of rice stands as an embodiment of the historical symbiosis and harmonious advancement shared between humanity and the rice plant.

Around a decade of millennia in the past, the Zhejiang province bore witness to a cultural epoch known as the "Shangshan Culture." Botanists meticulously sieved through remnants at this site, revealing a mere duo of carbonized rice grains. Intriguingly, among the ancient pottery crafting practices, discernible traces of rice husks blended within the clay emerged. The intentional incorporation of rice husks into the clay medium likely served to mitigate the clay's viscosity, thus averting pottery fragility during the firing process. In light of their ceramic craftsmanship and the vestiges of incinerated rice husks, it is plausible to infer that rice had become an integral facet of ancient existence in the region, eons ago. Their endeavors extended beyond rice grain collection to the creative application of rice husks.

However, definitive traces of agricultural tools remained absent from the Shangshan site. Noteworthy are the substantial stone fragments discovered there, possibly moonlighting as stone knives or sickles for reaping wild rice within the verdant fields. It seems conceivable that early humans commenced rudimentary rice cultivation efforts, albeit in a primitive state.

Subsequent to the Shangshan era, some seven to six millennia ago, the Hemudu Culture burgeoned in the fertile plains of Eastern Zhejiang. Within assorted sites, botanists unearthed an abundant array of botanical remnants, among which resided rice grains. Through meticulous botanical scrutiny, it was unveiled that during the nascent phase of the Hemudu Culture, roughly 6,900 years prior, cultivated rice accounted for less than 30% of the culinary mosaic. By the juncture around 6,600 years ago, the proportion of cultivated rice ascended to nearly 40%. This archival insight yields a narrative of arduous attempts by ancestral inhabitants to elevate their rice cultivation finesse. Nonetheless, it remains evident that during this interval, cultivated rice had not yet eclipsed the dominion of its wild precursor.

Within this timeframe, a panoply of rivaling botanical contenders vied for sustenance supremacy alongside rice. Archaeological surveys of Hemudu Culture sites unveiled an assortment of other palatable wild flora, including water caltrop, oak fruits, prickly water lilies, persimmons, and kiwi. The oak fruits is notably endowed with ample starch and efficient storability, and it is worth highlighting that storage pits replete with oak fruits were discernible within these archaeological enclaves. Hence, these relics illuminate a bygone epoch exceeding six millennia, characterizing rice agriculture as a rudimentary fixture, subsisting as one among various culinary staples, its yield modest in comparison.

When, one may ponder, did the vast expanse of China truly embrace the identity of a rice society?

In Yuhang, Zhejiang, the magnificent Liangzhu Culture city is presented before the world. This enormous ancient city consists of three layers: the outer layer, covering 8 square kilometers; the inner city, covering around 3 square kilometers, with a wall circumference of 6.8 kilometers and a wall base width ranging from 20 to 145 meters, constructed entirely from stacked large pebbles, with walls made from packed yellow soil; the center of the inner city is the palace, with a palace foundation that is a well-ordered rectangular raised platform, about 10 meters high, 670 meters long from east to west, and 450 meters wide from north to south. To the north and northwest outside the city, ancient people constructed large-scale water management systems using 11 mud-wrapped grass embankments to protect the city from flooding.

This urban tableau unveils a bygone civilization at its zenith—the venerable Liangzhu Culture—ensconced within the domains of Jiangsu and Zhejiang provinces.

The Liangzhu Culture was widespread in the area around the Taihu Lake, dating back to 5,200 to 4,300 years ago. Archaeologists have noted a significant increase in cultural sites from that period in the Jiangsu and Zhejiang regions, reflecting a substantial population growth. Archaeologists estimate that the construction of the central high platform of the ancient city on Mojiao mountain, the inner city walls, the outer city walls, and the surrounding large-scale water management systems of Liangzhu requires around 12 million cubic meters of earthworks. If calculated simply at 1 cubic meter of earthwork per person per day, approximately 33,000 people would need to work continuously for a year. Calculated at 10,000 people working 200 days a year, the entire construction project would take over six years.

To orchestrate such colossal endeavors within an antiquated society, some five millennia ago, is indicative of the exigency for an ample workforce and the substantial provisioning of sustenance and essentials. This would entail the harmonious collaboration of a cadre of agrarian and artisanal hands.

Indeed, only the vanguard of agricultural prowess could undergird the ascent of Liangzhu Culture and the coronation of Liangzhu's ancient citadel in the Orient. The bedrock upon which Liangzhu Culture stood was none other than the agrarian symphony of rice cultivation.

The new discoveries in plant archaeology confirm the rapid development of rice agriculture during the Liangzhu Culture Period. For instance, the Maoshan site, belonging to the Liangzhu Culture, features ancient rice paddies divided into elongated plots by irrigation channels and embankments. Each plot ranges from 1,000 to 2,000 square meters, with a total area of ancient rice paddies covering 56,000 square meters, equivalent to 84 acres. Additionally, a large ash pit was discovered on the edge of a high platform at Mojiao mountain. A remarkable amount of charred rice was excavated from the pit, suggesting it was originally a grain storage pit that had caught fire. Calculations indicate that the total weight of charred rice in the pit prior to carbonization was around 13 tons. Based on its location, this storage pit might have belonged to upper-class individuals, making the fire incident a significant loss.

The zenith of Liangzhu's cultural eminence bestrides the historical stage as a harbinger of an epoch wherein, following an intricate ballet of selective rice breeding and cultivation, the annals of time capitulated some five millennia ago to embrace the dawn of the rice cultivation era. Within the symphony of this societal tapestry, rice took the lead role as sustenance par excellence. As a culinary lynchpin, it harmonized its essence with a symposium of other botanical and faunal elements to consummate a culinary idyll.

Rice originated in Southern China and spread soutward from there, reaching Southeast Asia and South Asia to the south, and being carried across the sea by early inhabitants to the Japanese Archipelago and the Korean Peninsula to the East. In fact, during the domestication process in the middle and lower reaches of the Yangtze River and the Qiantang River Basin, rice began to spread in all directions. Archaeological evidence shows that rice cultivation had already moved beyond the Yellow River Basin and settled in areas downstream of the Yellow River at least 7,000 years ago. By 5,600 years ago, rice had even entered the Shaanxi Guanzhong region, which was dominated by northern crops like proso millet and foxtail millet, leading to a mixed cultivation pattern of rice, millet, and other crops.

Within this symphonic odyssey, one cannot overlook the role of rice as a culinary vestige of antiquity, a sustenance tethered to the precincts of the Jiahu archaeological site. Euphonic excavations herald an assemblage of carbonized botanical embryos, among which rice grains luxuriate, exceeding a sumptuous tally of 400. This compilation hints at the cerebral forays of Jiahu's denizens in the realm of southern Central Plains—a matrix where they gleaned insights from the confluence of the Yangtze's tributaries and embarked upon the nascent venture of rice cultivation.

As ancient people from 8,000 to 9,000 years ago, the way the Jiahu people acquired food was somewhat similar to the Hemudu people in the south. The Jiahu site yielded numerous remains of caltrop, lotus root, oak fruits, soybeans, and more. However, the Jiahu people's ability to leisurely play bone flutes could also be attributed to their remarkable fishing and hunting skills. The site also contained a substantial number of fish bones and shell fragments, indicating a well-developed fishing industry. Considering that their plant-based foods included aquatic resources like lotus root and caltrop, it can be inferred that the Jiahu people had a lifestyle heavily reliant on water. Their environment likely featured extensive water bodies, providing abundant aquatic resources.

Let us not omit the harmony resounding from the Jiahu's bone flutes, fabricated from the feathers of the red-crowned crane, a celestial choreographer known to entwine its elegant raiment with rippling waters. In retrospect, one can envisage the existence of the Jiahu community as one of pastoral contentment, where the rhythms of sustenance often orchestrated the symphony of abundance. This affluence not only allowed them to realize the quote that "abundant granaries lead to understanding of etiquette," empowering them to indulge in the art of flute playing, but also enabled them to raise livestock, such as pigs, another food source with ancient origins in China.

3.3 Swines and Canines: Ancient Humans' Best (Eating) Friends

The earliest remains of domesticated pig bones in ancient China were unearthed at the Jiahu site, possibly dating back to around 8,500 years ago. It appears that the people of Jiahu began raising and consuming pigs around that time. The transition from hunting wild boars in the natural environment to raising domestic pigs was a lengthy process, involving several key stages for ancient humans to overcome.

The domestication of wild boars had its genesis in the inherent yearning for meat. The ancestral populace, long reliant upon the art of hunting for their sustenance, had become acclimated to indulging in the pleasures of flesh, predominantly sourced from deer and wild boars. However, the burgeoning populace and protracted habitation within specific territories rendered the erstwhile bountiful hunting grounds barren, necessitating novel avenues for procuring meat. Hence, it was during this juncture that the ancients embarked upon capturing young wild boars, ushering them into maturation before eventually succumbing to the blade for consumption.

Furthermore, ancient humans needed a sufficient quantity of food. Apart from meeting their own nutritional needs, they required surplus resources to feed domesticated pigs. This implied that their agricultural techniques had reached an advanced level, capable of consistently yielding substantial crops. Therefore, the domestication of pigs likely emerged after a certain level of agricultural development. Additionally, pigs are omnivorous animals with undemanding dietary requirements, which contributed to ancient humans' ability to domesticate them relatively early.

Both archaeological and genetic evidence indicate that ancient people in East Asia engaged in the domestication of wild boars, possibly on more than one occasion. In the northern regions, represented by the domestic pigs from the Jiahu site, and in the southern regions, represented by domestic pigs from slightly later sites like Kuahuqiao, it appears that domestic pig husbandry emerged in the Yellow River and Yangtze River Basins around 8,000 years ago.

Animal scientists have collected genetic samples from ancient pigs in the Yellow River region and analyzed their relationships. The results suggest that the ancient pigs in the Yellow River Basin likely originated from a single domestication center. Considering that pig husbandry appeared later in the upstream areas of the Yellow River compared to the middle and lower reaches, it can be inferred that the origin of northern domestic pigs was likely in the middle and lower reaches of the Yellow River, including sites like Jiahu. Following the domestication of pigs more than 8,000 years ago, pig farming gained momentum in the Yellow River Basin. By over 6,000 years ago, pork constituted an increasingly significant portion of ancient humans' dietary resources, even accounting for up to 80%.

Raising pigs was a complex practice. The knowledge of pig husbandry in northern regions was continuously accumulated over time. Analyses of pig bones from archaeological sites revealed that around 8,000 years ago, slaughtered pigs were likely around six months to a year old, indicating that ancient humans were eager to consume younger pigs. By 4,500 years ago, the slaughtered pigs were around a year old or older. Under the prevailing feeding conditions, the highest meat yield was achieved when pigs were raised to a year old. Younger pigs put on minimal weight, while older pigs' growth slowed, both scenarios resulting in inefficient feed conversion.

A pertinent inquiry surfaces: did the domestic swine of that era engage in a subtle "rivalry for sustenance" with their human custodians? Substantiation emerges from the intricacies of carbon isotope compositions encapsulated within their ossified remains. Ancient society was reliant on gathered sustenance from the plant kingdom, often comprising of the categorically designated C3 plants. Conversely, crops meticulously cultivated, such as foxtail millet and proso millet, hailed from the C4 lineage. The symphony of antiquity resonates from a bygone era 4,000 years distant: human and swine remains coalesce in testament. An intricate analysis discloses a diet heavily steeped in C4 plants for both human and porcine palates. This confluence underscores a shared predilection for millet-based fare—humans indulging in millet and millet-borne libations, while swine partaking of millet husks and the vestiges of vinous fermentation.

However, in the Yangtze River Basin, domestic pigs did not receive the same enthusiasm from ancient people. Although people in the southern region began domesticating pigs over 7,000 years ago, it seems that they raised pigs sporadically without considering it a significant part of their lives. In the archaeological sites of ancient Southern China dating back more than 7,000 to 5,000 years ago, the number of domestic pig remains unearthed was consistently low. Their meat consumption primarily relied on hunting wild animals like deer, or fishing in rivers and lakes for fish and shellfish. In a site dated around 5,500 to 5,100 years ago in the Three Gorges area of the Yangtze River, a female burial was found with two large fish placed under her arms, each fish about half a meter long, nearly the same length as the woman's arms. This practice of burying fish as offerings was not uncommon, indicating the greater emphasis on fish resources among the ancient people of that time. The raising of domestic pigs might have been driven by a diverse diet and not solely due to a lack of meat.

The formation of the ancient Chinese dietary pattern known as "fish in the south, pork in the north" more than 5,000 years ago was likely influenced by the natural environment. The northern region had fewer lakes and rivers compared to the south, and these water bodies would freeze during winter. As a result, the Yellow River Basin had fewer fish resources compared to the Yangtze River Basin. Another possible factor was that ancient people in the north successfully domesticated millet crops such as proso millet and foxtail millet, leading to surplus grain production. In contrast, during the same period, the domestication of rice in the southern region was still underway, leaving less surplus grain. This, in turn, influenced the development of pig farming in the south.

However, the "fish in the south, pork in the north" dietary pattern underwent a significant change with the rise of the Liangzhu Culture. In Liangzhu Culture archaeological sites, domestic pig bones suddenly dominated the animal remains found, indicating a prolonged and large-scale pig farming activity. The thriving rice agriculture of the Liangzhu Culture provided the foundation for pig feed. The enthusiasm for pig farming among the Liangzhu people suggests a high population concentration near their settlements. Given the extensive construction efforts evident in the monumental Liangzhu ancient city, it can be inferred that a large number of managers and laborers congregated for an extended period. The surrounding environment could no longer sustain their meat needs through fishing and shellfish gathering alone, prompting them to turn to pig farming.

In ancient times, the Liangzhu Culture was an exception in the southern region, representing a prehistoric leap forward in a wetland environment. The Liangzhu people maximized rice cultivation to secure abundant food, support a large population, and raise domestic pigs for improved nutrition, all of which were crucial for their massive construction projects.

Their zeal for constructing cities, houses, and water systems was tied to their mysterious beliefs. Archaeologists sense a fervent devotion among the Liangzhu people to a unique faith that drove them not to establish a dominating "kingdom," but to dream of building an earthly "paradise." This faith compelled them to transcend their era and achieve unparalleled greatness.

Thus, the Liangzhu Culture carries a sense of grandeur as a period with no equal before or after. From the perspective of the subsequent Maqiao Culture that arose after the decline of the Liangzhu Culture, the ancient people of the Southern Yangtze River Basin and Qiantang River Basin reverted to a cultural state reminiscent of the time before Liangzhu, relying on fishing and hunting for meat. They almost ceased pig farming and the intricate carving of exquisite jade artifacts. Their social structure became significantly simpler than during the Liangzhu Culture era, as if the Liangzhu Period had never occurred. A revival of a cultural peak moment in the southern region might not occur until the Wu and Yue power struggle in the Spring and Autumn Period, around 500 BCE, approximately 1,800 years after the decline of the Liangzhu Culture.

Leaving Liangzhu behind, let's broaden our view globally and further understand the intriguing connection between pigs and humans. Genetic research on pigs suggests that the ancient relationship between the people of China and pigs might date back even earlier than the Jiahu site. Genetic analysis of domesticated and wild pigs across various regions in Asia indicates that all these pigs may share a common ancestor, one that was domesticated in the Mekong River Basin of Southeast Asia. Subsequently, these pigs spread in two directions: northwest and northeast, eventually entering the territory of present-day China. The dispersal route of pigs is quite similar to the path we described in the previous chapter for the entry of modern humans into the Chinese land, both branching out northward from coastal and inland areas. Therefore, the initial domesticated pigs in the Chinese land were likely brought by modern humans from Southeast Asia.

It's important to note that the differences between wild and domesticated pigs around the world are not significant, and they can interbreed. According to biological definitions, strictly speaking, animals capable of interbreeding to produce fertile offspring should not be classified as separate species. Thus, the distinction between domesticated and wild pigs can only be seen as a difference within a single species—domesticated pigs are essentially less wild "wild pigs." As modern humans migrated with their domesticated pigs, these pigs would revert to a wild state in the wild and interbreed with wild pigs along the way, gradually reducing the wildness of the latter. Such complex hybridization scenarios have made it difficult for future biologists to definitively determine where domesticated pigs were initially tamed. Although they generally believe there might be six domestication sites for pigs worldwide, the Yangtze River Basin in China and the Mekong River Basin in Southeast Asia are among them.

Thus, the evolutionary process of domesticated pigs somewhat aligns with the hypothesis of continuous human evolution accompanied by hybridization, as discussed in the first chapter. In summary, European domesticated pigs likely descended from European wild boars, while Asian domesticated pigs originated from Asian wild boars. However, an exception lies in the genetic analysis of ancient pigs found in Guangxi, China, which indicates a closer relationship to European ancient pigs than to Asian ones. The story of humans and pigs is indeed quite convoluted.

Setting aside these complex relationships with pigs, let's turn our attention to another animal with a relatively straightforward origin from China—dogs.

Dogs are among humanity's companions, clearly having been domesticated from wolves in the natural world. These two animals are quite similar, yet there are distinct genetic differences between them. Both dogs and wolves enjoy meat; offer them pork, and they'll happily wag their tails and devour it. Give them a potato, and while wolves might disdainfully ignore the starchy food, dogs will readily consume it. This isn't because wolves are too proud to eat potatoes, but rather because their genes prevent them from digesting starch.

By comparing the complete genomes of dogs and wolves, scientists have identified at least three genes related to starch digestion in the dog genome, which are absent in wolves. One of these genes encodes an enzyme that breaks down starch into sugars. Without this enzyme, animals can't digest starch.

Hence, it's reasonable to speculate that only those wolves that evolved genes capable of digesting starch eventually became dogs. Following this clue allows us to unveil the story of the human–dog connection.

Scientists from both China and other countries collected genetic samples from over 1,500 dogs across Asia, Africa, and Europe to explore the origin of dogs. By comparing genes, they found that the dog species likely originated in the southern part of China's Yangtze River Basin, around 16,000 years ago. These particular wolves might have been drawn to human habitation by gathered plant-based foods, which led to their capture by early humans. The earliest domesticated dogs weren't meant for hunting or guarding, but as a source of food—they were initially bred for consumption.

Kind-hearted dog owners today shouldn't be angered by this phenomenon; it's worth considering that if our hungry ancestors had perished prematurely due to lack of sustenance, we wouldn't be here as their descendants. As rice cultivation began to flourish thousands of years ago in the southern regions, ancient humans had more plant-based foods to feed dogs, leading to more dogs being consumed and consequently more dogs coexisting with humans on Earth.

Archaeologists have unearthed the earliest remains of domesticated dogs within central Hebei province, China, dating back ten millennia. The Jiahu site has also yielded large dog remains, somewhat supporting the notion that dogs were initially regarded as a food source—ancient people selected and consumed meatier dogs.

As a native-origin animal, dogs were widespread in ancient China and were present across vast lands from the northeastern regions to the Qinghai-Xizang Plateau over 5,000 years ago. As their relationship with humans deepened, dogs became more docile, serving not only as sources of meat but also as guards for households. The fate of dogs in human society was gradually changing.

3.4 Agricultural Paradise: Where I Find My Place

Proso millet, foxtail millet, beans, rice, pigs, dogs ... After the end of the Younger Dryas cold period, the ancestors of Huaxia seemed to have "hacked" into the beginning of the agricultural era, delivering a series of "aces" right from the start. They quickly established the foundation for the agricultural society that would shape Chinese civilization for thousands of years to come. There is no other region of comparable size in the world that can rival the ancient agricultural scale and technology of the land of China.

This agricultural paradise emerged in the human world, in the land of China, thanks to the synergy of opportune timing (天时), geographic advantage (地利), and unity among all ranks (人和).

The term "opportune timing" refers to the "Yangshao Warm Period," internationally known as the Holocene Climatic Optimum. This period corresponds to the pre-Yangshao cultural period, Yangshao Culture, Longshan Culture, and Xia Culture in Chinese archaeology. Scholars use "Yangshao" to describe this warm period in China; however, the warm period lasted much longer than the Yangshao Culture Period. The ancestors, having endured the cold, encountered a prolonged period of warm climate, allowing them to accumulate and exchange agricultural technologies and species with ease.

During the 5,000 years from 8,000 to 3,000 years ago, the world basked in warm sunlight. Although there were occasional cooling periods, they were brief and followed by a return to warmth. Moreover, China was one of the regions that experienced the highest temperature increase during this global warm period. Based on plant pollen deposits in the Loess Plateau, Hebei, Qinghai, Inner Mongolia, and Liaoning regions, climatologists estimate that the average temperature in these areas was 2 to 4 degrees Celsius higher 8,500 to 7,200 years ago compared to today. The subsequent period from 7,200 to 6,000 years ago was a stable warm and humid phase within the Yangshao Warm Period, characterized by generally warm and humid climates across the country, with monsoon rains reaching all corners and vegetation thriving like never before.

The term "geographic advantage" refers to the suitable farming conditions of the Loess Plateau. The Loess Plateau spans provinces and autonomous regions such as Shanxi, Shaanxi, Gansu, and Ningxia in modern China, covering an area of over 400,000 square kilometers. It is covered by a layer of loess soil several meters deep, which has been blown in by westerly winds over the past two to three million years. Due to the uplift of the Qinghai-Xizang Plateau blocking the penetration of the Southeast monsoon into the interior, Western China and Central Asia became arid. These areas also happen to lie in the westerly wind belt of the Northern Hemisphere. As a result, the westerly winds carry fine sand and dust eastward, depositing them in the present-day location of the Loess Plateau. Over time, this accumulation formed the Loess Plateau.

Loess soil is uniform, fine-grained, loose, and friable, with high organic matter content and good water retention properties. It has a "self-fertilizing" capability, making it a precious land for the initial Chinese farmers. Initially, slash-and-burn agriculture was practiced, where farmers cultivated a piece of land for a year before leaving it fallow and moving on to another plot to avoid soil fertility decline and degradation. However, the Loess Plateau's soil maintained its fertility as long as proper soil-water management was upheld. Moreover, during the Yangshao Warm Period's onset, when temperatures were higher than today, the natural environment on the Loess Plateau was excellent, with abundant vegetation covering the land surface, and soil erosion was not severe. This environment was highly suitable for the continuous cultivation of dryland crops such as millet and beans.

The term "unity among all ranks" signifies our ancestors' wisdom in innovating tools and domesticating various wild species. We have already seen their domestication of plants and animals, which accompanied the advancement of production tools. At the beginning of agriculture, they used traditional stone axes to fell trees, stone shovels to till the soil, stone sickles to harvest wild and domesticated crops, and stone querns to grind fruits and seeds collected. These tools were carried over from the fishing, hunting, and gathering ages. To adapt to the needs of agriculture, they improved stone tools, crafting oval axes, perforated axes, and compound tools like perforated shovels. They also invented new digging tools known as "leisi" (耒耜), which evolved from pointed wooden sticks used in digging plant roots during the gathering age. Later, it developed into a shovel-like implement, becoming one of the most popular agricultural tools in ancient China. Until the Bronze Age arrived, leisi remained an important tool for ancient Chinese farmers.

With the synergy of opportune timing, geographic advantage, and unity among all ranks, the numerous granaries unearthed from sites across the Yangshao Warm Period tell us that our ancestors were harvesting more and more grains. They no longer roamed around for food like in the fishing, hunting, and gathering ages; they could settle down for extended periods, enjoying a harmonious life in their own corner of fertile land. In comparison to the fishing, hunting, and gathering ages, the population on the land of China exploded during the Yangshao Warm Period.

Scholars have once estimated the population scale of the Yellow River middle and lower reaches during the Yaoshao Warm Period. This region roughly includes the entire or a significant part of what is now Henan, Shanxi, Shandong, and Shaanxi provinces. Around 8,500 to 7,000 years ago, when crops like millet and barley, as well as animals like pigs and dogs, had just been domesticated, the population in the Yellow River middle and lower reaches was likely around 200,000 to 300,000. Of course, this population estimate might already be an order of magnitude higher than during the fishing and hunting gathering era.

Around 7,000 to 5,000 years ago, during the most warm and humid phase of the Yaoshao Warm Period, the population in the Yellow River middle and lower reaches might have first exceeded the one million mark and then gradually increased to around two million, representing another order of magnitude increase compared to before. During this period, the Yangshao Culture, marked by its painted pottery, began to flourish in Northern China.

In the following period, from around 5,000 to 4,000 years ago, dryland agriculture in the northern regions had become quite advanced. The population in the Yellow River middle and lower reaches doubled again, potentially reaching a scale of four to seven million. During this time, it seems that cultures from several directions radiated toward the Central Plains. The Hongshan Culture, based in the northeast, spread across the Yanshan Mountains into the Central Plains; cultures from the central part of Inner Mongolia to the north crossed the Yinshan Mountains to reach the Central Plains; the Yangshao Culture, originating from the west in Shaanxi, spread across the Yellow River to the Central Plains.

This convergence of cultures led scholars to draw parallels with the ancient Chinese myth of the battle between the Yellow Emperor (黄帝, Huáng Dì), Yan Emperor (炎帝, Yán Dì), and Chiyou (蚩尤). Regardless of how much the ancient myths align with actual ancient cultures, the cultural landscape that emerged in Northern China during this time can be described as a struggle for dominance in the Central Plains.

The participants in this struggle for dominance were not so much diverse groups of people, but rather various ancient cultures from different regions. With the onset of agriculture, the population on the land increased by at least two orders of magnitude. Many settled communities emerged with high population densities, fostering prosperous settlements. If the period before agriculture was characterized by genetic flow, then this era saw a greater emphasis on cultural exchange. The struggle for dominance wasn't just about people with different genetic backgrounds converging in the Central Plains; it was about ancient cultures from different areas radiating outward, including into the Central Plains. This diffusion was facilitated by interactions among populations, involving the exchange of cultural and technological knowledge through gifting, trade, and other means.

In contrast to the explosive population growth and cultural convergence in the north, the southern regions appeared to lag behind. Apart from the Liangzhu Culture, which was an exception as mentioned earlier, the overall development in the southern regions seemed to progress at a slower pace, resembling a notion of "small states and few people" (小邦寡民).

In the era of genetic shift, the southern regions were once vibrant pioneers, serving as cradles where diverse populations carrying different genes accumulated strength and as starting points for expansion northward. However, why did the trend of southern strength and northern weakness reverse after the onset of the agricultural era, transforming into a situation of southern weakness and northern strength?

The resolution to this riddle finds its roots embedded within the geographical context. Notably, the southern climes, with their elevated thermal regimes vis-à-vis their northern counterparts, basked in an environment replete with biodiversity, relatively insulated from the perturbations of Ice Ages. Thus, in the antecedent epochs, the southern landscapes, endowed with ecological abundance, poised as ideal niches for the sustenance pursuits of ancestral societies, distinguished by maritime and terrestrial harvests. With the emergence of agriculture, particularly the development of dryland farming in the northern areas using crops like millet and beans, the survival advantage of the southern regions relative to the north disappeared.

Even though the southern regions possessed rice through domestication, their agricultural practices were notably disadvantaged compared to the north, which had easily cultivable loess soils and extensive plains for farming. For instance, the southern soil was predominantly compact clay, and abundant vegetation covered the land. Before the Bronze Age, clearing southern fields using stone tools was more challenging than in the north. The presence of dense mountain ranges and crisscrossing river networks further fragmented the limited arable plains. The high rainfall often led to flooding disasters. The lack of contiguous farming areas as a material foundation meant that the culture nurtured in the south couldn't match the scale of the northern culture. The high yield of rice fields demanded advanced agricultural techniques, from seedling cultivation to harvest. Southern farmers needed to master a series of skills, such as leveling ponds, constructing water channels, and maintaining field embankments.

A resurgence of the south relative to the north would likely take several hundred years after the Common Era, at the very least.

The advent of the agricultural epoch heralded two distinct genetic shifts among ancient societies.

Foremost, the genes of the agrarian populace gradually ascended to prominence within the annals of the human genetic repository. A burgeoning populace nourished by agrarian practices orchestrated a gradual inversion of fortunes, where the prevailing dominion of hunter-gatherer genotypic expressions ceded territory to an ascending prominence of agrarian genotypic propensities. This transformation underscored the transcendence of farmer-centric genetic paradigms to the zenith of global prevalence, relegating the once-dominant hunter-gatherer genotypes to the fringes of the genetic spectrum, confined to the realm of the niche and the periphery.

Secondary to this genetic realignment was the remarkable observation that despite the expansion of agrarian societies, the exchange of genetic information among diverse groups did not exhibit a proportional surge, particularly during the nascent phases of the agrarian era. The migratory patterns of erstwhile free-ranging hunter-gatherers, now bowed in diligent cultivation as agrarians, led to the gradual consolidation of sedentary lifestyles for numerous ancient groups. The quotidian pursuits of these populations revolved around their allotted plots of land, constricting interactions primarily within the bounds of their tribal or village confines. The result was a discernible state of "reproductive isolation," a phenomenon that dictated the preservation of unique genetic characteristics within disparate agrarian communities during the Yangshao Warm Period. The melding of genes on a grand scale would, with the passage of time, bestow its influence upon the expansive terrain of China.

Gazing back across the chasm of over four millennia to this pivotal crossroads of history, one finds that the coalescence of cultures far outshone the ebb and flow of genetic exchanges. Across the northern expanse, myriad cultures intertwined and vied for ascendency, akin to a constellation of radiant stars adorning the historical firmament. Imminently, the relentless winds from the north were poised to surge forth, heralding a seismic transformation reminiscent of the phoenix's resplendent rebirth, destined to sweep across the vast tapestry of China's landscape. Our forebears stood on the precipice of embracing an epoch anew.

The Column: How Humans Initiated Agriculture

In ancient times, how much land was needed to feed a family of five? It depended on how they obtained their food.

Relying on hunting for food, about 0.001 to 0.05 people could be supported per square kilometer of land. In other words, a family of five engaged in hunting would need a "hunting ground" of about 100 square kilometers. On the other hand, relying on agriculture, there was a saying in ancient China: "Thirty mu (亩, unit of area) of land can support a family of five with a cow." Thirty mu of land is roughly equivalent to 0.02 square kilometers. This means that the land required for an agricultural population is only a tiny fraction of what hunting populations need. Agriculture greatly increased the amount of food that humans could obtain.

While agriculture had its advantages, not all ancient communities were eager to become farmers. After all, hunting and gathering weren't particularly arduous ways of subsistence. Hunter-gatherers didn't need to work too hard, nor did they need to raise many children as helpers. By migrating to resource-rich areas, they could lead a decent life. Becoming a farmer meant settling down in one place, raising more children, engaging in labor-intensive work, and dealing with the disruptions of natural disasters like droughts and floods.

So, why did some ancient people choose to become farmers? They might have been compelled to do so. At a certain point, hunter-gatherers lost the ability to lead migratory lives, or they gave up migratory lifestyles for specific reasons.

Around 12,000 years ago, Earth experienced a cold period known as the Younger Dryas. During this time, both animal and plant resources significantly decreased. In the region of West Asia, many tribes found it difficult to migrate long distances due to the cold winds, forcing them to settle in one place and produce their own sustenance. Along the coast of the West Asian Mediterranean, there existed the Natufian Culture. These ancient people repeatedly attempted to domesticate and cultivate wild grains, hoping to harvest seeds from these plants to supplement their scarce food. As the cold period persisted, they began domesticating and planting wild grains each year. This marked the beginning of agriculture, as the ancient people who had sheltered from the cold in these regions painstakingly explored new modes of production.

There are three main centers of origin for world agriculture: West Asia–North Africa, East Asia, and the Mexico–South America Andes region. These areas weren't necessarily the most resource-rich in terms of flora and fauna; rather, they had semi-humid to semi-arid climates. Under the pressure of food scarcity during the Ice Age, ancient communities that were thousands of miles apart coincidentally transitioned from hunting and gathering to cultivating plants and animals. Although cultivation required waiting for harvest and the yield wasn't as nutritious as hunted and gathered food, at least the harvest was relatively stable. A bit of cultivation led to a bit of harvest.

These resilient ancient people survived the cold Ice Age and entered the agricultural era of prosperity.

In Western Asia, newly cultivating farmers continuously exchanged various technologies, knowledge, seeds, and livestock-raising experiences with neighboring tribes. Within a few hundred years, the number of farming villages in plains and foothills with ample water sources kept increasing. Centers emerged for various improved varieties of wheat, barley, goats, and sheep. Many other types of crops were also quickly incorporated into ancient people's domestication plans. Various legumes began to appear as supplementary crops. In suitable conditions, olives, grapes, figs, and dates became valuable crops. A variety of green vegetables and spices added diversity and nutrition to human diets. Humans also extracted fiber from flax plants for weaving.

Across the globe, ancient people in Northern and Southern China domesticated foxtail millet, proso millet, soybeans, and rice. In the Americas, ancient communities domesticated corn, sweet potatoes, potatoes, peanuts, tobacco, peppers, sunflowers, pumpkins, and lima beans, among others. Additionally, ancient Indians domesticated cotton, while Southeast Asians domesticated taro. People south of the Sahara in Africa domesticated sorghum, pearl millet, African rice, and more.

Furthermore, once a certain crop was domesticated, ancient people quickly spread it to suitable planting areas far away, unless there were significant geographical barriers. This was because, when agriculture first emerged, there were many fertile virgin lands yet to be cultivated on Earth. The yield of the first crop planting would be high, but if planted continuously for years, soil fertility would decline and compacted soil would be difficult to cultivate. Therefore, our farming ancestors would plant crops on one piece of land for a season, then move to another piece of land nearby that hadn't been cultivated yet, leaving the previously cultivated land fallow.

Hence, the earliest post-Ice Age farmers might not have engaged in purely settled agriculture but rather practiced migratory agriculture, "shooting one arrow and moving to another place." This facilitated the rapid spread of certain crops to distant areas. They also set fire to clear land, burning down forests to reveal fertile soil for cultivation. It is estimated that the early farmers had about nine times the cultivated land per person compared to modern farmers. Of course, the population of farmers back then was very small, but their widespread practice of burning and cultivating released a significant amount of carbon dioxide, possibly leading to a notable increase in atmospheric carbon dioxide around 6,000 years ago.

Certain scientists speculate that following the Ice Age, the Earth's environment transitioned into an extended epoch of comfortable warmth spanning several millennia. This phenomenon is attributed, in part, to natural forces and, conversely, to the deliberate actions of humanity, particularly in terms of controlled burning for land clearance. The elevation of atmospheric carbon dioxide, in a reciprocal manner, has proven conducive to amplifying agricultural outputs. Empirical investigations have revealed that heightened carbon dioxide concentrations correlate with enhanced yields across a plethora of plant species, with the most pronounced effects observed in the case of uncultivated wheat and barley. This dynamic establishes a self-reinforcing cycle—our ancestors exercised preferential selection of prolific wild crops for cultivation, while their pioneering land clearance endeavors further augmented the productivity of domesticated crops.

The trajectory of agricultural inception and expansion, though seemingly auspicious, was underscored by a spectrum of challenges that the ancient populace confronted, notably among these, pathogenic ailments.

Sedentary agricultural communities found themselves considerably vulnerable to infections from pathogens and bacteria, as their mobility diminished vis-à-vis their hunter-gatherer forebears. Noteworthy is the close interaction that individuals had with the diverse waste byproducts arising from their burgeoning societies, engendering an environment of heightened contamination. In departing from the pathogen-rich African tropical domains to inhabit the climatically temperate realms of Eurasia, modern *Homo sapiens* initially experienced a mitigation of pathogenic threats. Paradoxically, this trend was reversed upon their transition from nomadic foragers to agrarian cultivators, as diseases once again gained prominence.

The process of animal domestication further exacerbated the human susceptibility to a myriad of maladies, as sustained proximity to livestock facilitated the zoonotic transmission of numerous ailments. A paradigmatic instance can be drawn from genetic analyses revealing that one of the most historically formidable pathogens to afflict humankind—smallpox—is believed to have originated from interactions with Arabian camels. In addition to this, the origin of various influenzas and the measles virus can be traced back to interactions with animals beyond the human sphere.

Perhaps most emblematic is the case of tuberculosis. Mycobacterium tuberculosis, the causative agent of human tuberculosis, has wrought considerable mortality on a global scale, with an annual toll numbering in the thousands. Genetic inquests suggest that the earliest strains of tuberculosis may have emerged in the African landscape tens of millennia ago; however, their impact on human society remained relatively subdued during that era. Paradoxically, the ancestor of contemporary tuberculosis bacteria emerged approximately ten millennia ago, aligning poignantly with the advent of agriculture. Geographical origins further accentuate this correlation, with tuberculosis progenitors arising in the fertile crescent of Western Asia and subsequently dispersing globally through the migratory and communicative networks of early agrarian societies. Thus, the trajectory of human agricultural advancement, while satiating the physiological needs of sustenance, concurrently underpinned the swift proliferation of tuberculosis pathogens.

CHAPTER 4

The Path of the Prairie: Enkindling the Broadband of East Asian Civilization Transmission

Behold the grandeur of a bygone era, where China's ancient history unfurls its secrets within the sprawling expanse of a city that surpasses all in antiquity, spanning over 4 million square meters. Even the contemporaneous magnificence of Liangzhu Culture's ancient citadel stands humbled in the shadow of this architectural marvel.

A trinity of urban magnificence graces this ancient citadel, adorned with a cityscape akin to a symphony in three movements—the outer city, the inner precincts, and the imperial sanctum, embraced by the square contours of its formidable city walls.

Exemplifying consummate architectural prowess, this ancient city beckons with walls bearing witness to a meticulous craft—the inner sanctum boasts walls molded from tamped earth, while the outer facade is a masterwork of stone masonry. Every surface, every corner, meticulously etched to perfection, a testament to the zenith of architectural finesse.

Within its stone embrace, this citadel reveals its indomitable spirit of defense, a bastion of impregnability. Intricacy defines its gateways, where a double-arched, fortified city gate stands as the sentinel between realms. Embellished along the wall's sinuous expanse, horse-faced bulwarks and sentinel towers stand as sentries, unwavering, vigilant, leaving no recess unchecked.

Thus emerges the legacy of the Shimao ancient city, a sentinel of the past, erected four millennia ago within the tapestry of Shenmu's northern hinterlands in Shaanxi province. In this colossal cityscape, fortified and resolute, one queries the architects of its grandeur and the shadow of conquest that compelled its vigilant ramparts to rise.

4.1 The Mystery of Shimao

The ancient city of Shimao is believed to have been established around 4,300 years ago and existed for about 300 years before it was abandoned. Located at the juncture of the Mongolian Plateau and the Loess Plateau, the ancient city lies in a mountainous region. To the north lies the vast green grasslands, while to the south lies the yellow, soft earth. The city's mysterious history is absent from historical records, so exploring the origins and details of the Shimao Culture relies solely on archaeological excavations.

The Shimao Culture shares deep cultural connections with the land of China, as evidenced by their reverence for jade artifacts. Various jade objects, such as jade shovels and jade rings, are embedded within the walls of the Shimao archaeological site. This isn't a whimsical act, but a deliberate arrangement where intricately carved jade pieces were inserted into the crevices between stone blocks. Jade artifacts have also been found as burial items in Shimao Culture tombs.

Early inhabitants of China fostered a jade culture, using jade artifacts to communicate with the divine. From the Hongshan Culture in the northeastern regions to the Liangzhu Culture in the southeastern areas, the reverence for jade has been a prominent tradition. It's clear that the Shimao Culture was deeply rooted in the cultural landscape of China.

Simultaneously, the Shimao Culture is imbued with elements of military and violence. Embedding jade artifacts into their city walls showcased their respect for jade culture, while also serving a clear purpose: invoking the divine through jade to safeguard the solidity of their walls. This action reflects a strong emphasis on military defense.

Several instances of painted geometric murals have been unearthed within the Shimao site, primarily on city walls and gates. These murals date back to different periods, likely corresponding to different phases of city wall renovations. The murals feature geometric patterns painted in colors like white, red, yellow, black, and green, and what's astonishing is the use of preliminary sketch lines. Preliminary sketching is an artistic technique where a light outline is drawn before the actual painting. Previously, the earliest known use of preliminary sketch lines in murals was from the Tang Dynasty. However, the murals discovered in the Shimao site date back to around 2,000 BCE, pushing back the emergence of this advanced painting technique by thousands of years.

Beyond artistic value, why did the ancient inhabitants of Shimao create painted murals on their city walls? Was it purely for aesthetic pleasure? Certainly not. The purpose of these murals, similar to the jade artifacts within the walls, likely served as a means for the ancient Shimao people to express certain desires, possibly related to enhancing their military defense.

Even more disturbing is the discovery of human sacrificial remains. Archaeologists found two locations near the eastern gate wall of the outer city of Shimao where 24 skulls were arranged in each site. Most of these skulls belonged to young females and displayed clear signs of cutting and burning. Archaeologists speculate that these tragic deaths might be linked to foundational or ritual activities during the construction of the city walls. Additionally, dozens of skulls were unearthed along the city's pathways, wall passages, and activity surfaces, presumably used for sacrificial purposes. Through these gruesome sacrificial rituals, the ancient Shimao people sought the everlasting protection of their city by appeasing the divine.

Around 4,000 years ago, the northern edge of the Loess Plateau was suitable for the cultivation of dryland crops like millet and beans. The advanced level of agriculture around the Shimao ancient city contributed to the prosperity of the Shimao Culture. This prosperity provided the resources necessary for the construction of massive city walls and the accumulation of wealth. The heavily fortified walls also indicate that the Shimao people faced considerable threats from neighboring powers, possibly leading to frequent conflicts and disturbances. Those sacrificed individuals likely originated from people captured during warfare.

Possessing significant strength and influence, the ancient Shimao people must have confronted formidable threats that kept them restless even within their secure walls. These threats might have emanated from all directions, but the most likely assailants who dared to launch sudden attacks on Shimao were likely from the northern grasslands.

4.2 Horses: From the Grasslands I Come

The horses domesticated by humans belong to the category of domestic horses, derived from the taming of wild horses in nature. Similar to the fate of dogs mentioned in the previous chapter, ancient humans initially regarded horses solely as a source of food. They either collectively hunted wild horses for sustenance or joined forces when other predators killed wild horses, driving away the predators and claiming the horse meat.

Archaeologists excavated a significant number of wild horse skeletal remains at an ancient human site in Northern Shanxi province, China, indicating that our northern ancestors hunted wild horses tens of thousands of years ago. During the Ice Age, they not only developed the ability to hunt wild horse herds on a large scale but also selectively targeted mature wild horses due to their higher fat and meat content.

With the advent of agriculture and settlement, ancient people, when food was abundant, began to raise the occasional captured young horse instead of immediately killing and consuming it. Over time, wild horses gradually evolved into domesticated horses.

The earliest domesticated horses likely emerged around 6,000–5,500 years ago in Central Asia. Archaeologists found ancient cooking utensils in present-day Northern Kazakhstan containing traces of mare's milk, along with several horse jawbones showing evidence of damage caused by reins. These findings indicate that ancient people of that time already had horse-raising ranches, consumed mare's milk extensively, and utilized the strength of powerful horses for labor.

Thus, domesticated horses likely originated from the taming of wild horses in Central Asia. A horse species called Przewalski's horse once lived on Chinese soil, with remains found in many ancient human sites ranging from Western Xinjiang province to the Taiwan Strait. Some scholars believed that early Chinese domesticated horses were derived from Przewalski's horse due to its historical geographic distribution and the frequent activities of prehistoric Eurasian grassland humans. However, mtDNA analysis between ancient, domesticated horses and Przewalski's horses indicated a distant genetic relationship, showing that Przewalski's horse is not the ancestor of ancient Chinese domesticated horses. The maternal gene sources of ancient Chinese domesticated horses were complex, suggesting their common ancestors existed in a very early era. Ultimately, the ancestors of Chinese domesticated horses also likely came from Central Asia.

As horses transitioned from the wild to domesticated state, their fate in human society was rewritten, and their status became much higher than that of dogs, which were domesticated from wolves. The pulling and carrying power of horses significantly aided ancient humans in transporting goods, enhancing their migratory capabilities. Even more exciting was the horse's ability to run at great speeds, allowing humans to surpass their physical limitations and travel swiftly to distant places. Initially, humans might not have been adept at riding horses and lacked the necessary equipment for riding, but they could at least cover long distances on horseback during migrations.

Starting from about 5,000 years ago, nomadic groups extensively began using horses for mobility, spreading in all directions across the Eurasian steppe and even venturing beyond it into the world of agrarian societies. For instance, in the Western Regions of the steppe, such as present-day Ukraine, where domesticated horses had just emerged, steppe communities began exchanging their livestock, animal hides, and wool with agrarian societies. Soon, these steppe groups intermittently raided agrarian villages on horseback, plundering their possessions.

The domestication of large herbivores like horses, cattle, and camels provided ancient human societies with a new way of life—pastoralism. People discovered that through domesticated animals, they could live on the grasslands, partially or completely detached from an agrarian lifestyle. Hence, pastoralism emerged from agrarian societies rather than directly evolving from ancient hunting and gathering practices. After all, the domestication of large animals was predicated on the accumulation of agricultural surplus.

Pastoralism primarily comes in two forms. One is pure pastoralism, where nomadic groups as a whole move their herds across the grasslands in search of water and grass. They stay at several pastures during the year, eventually returning to the starting point. With the assistance of large animals, their range of activity can become vast, spanning hundreds to thousands of kilometers. From the Mongolian Plateau to the Central Asian steppe, and further to the grasslands of West Asia and North Africa, numerous pastoral communities historically engaged in pure nomadic lives.

The other form of pastoralism is seasonal pastoralism, where some members of the community remain in agrarian villages while others tend to livestock. Typically, during a specific season like summer, they drive their animals to highland pastures. Pastoralism is just one of their means of livelihood; they also participate in agriculture, gathering, and other supplementary activities.

Let's shift our focus back to the Huaxia region. In the vast land of China, located at the eastern end of the Eurasian continent, the appearance of domestic horses seems to have occurred relatively late and suddenly. Archaeologists have long discovered that domestic horses and horse-drawn carriages suddenly emerged during the late Shang Dynasty, around 3,000 years ago. Many horse pits and chariot pits used for burial and ritual purposes were unearthed in sites from the late Shang Dynasty, including Anyang Yin Ruins in Henan, Shaanxi, Shandong provinces, and other regions. Horse remains were also found within them. In the period before the late Shang Dynasty, there were very few horse remains unearthed in sites dating back to 3,300 to 4,000 years ago. The earlier horse remains from ancient times primarily consisted of wild horses used as food, not domesticated horses.

Moreover, in the western parts of the Eurasian continent, the evolution of horse-drawn carriages has a relatively complete developmental process, transitioning from four-wheeled to two-wheeled carriages, from solid wheels to spoked wheels, and from ox-drawn to horse-drawn carriages. However, the horse-drawn carriages that appeared during the late Shang Dynasty were already sophisticated, directly presenting as two-wheeled carriages with spokes.

Various signs indicate that there was a lack of gradual domestication of horses and a process of continuous evolution and improvement of horse-drawn carriages within the Huaxia region. This once again demonstrates that the domestication of horses and the invention of horse-drawn carriages did not originate within the Huaxia region; the domestic horses and carriages that appeared during the late Shang Dynasty were likely introduced from external sources.

If we expand our perspective to include the northern grasslands within the Huaxia region, we might find that the northern Chinese grasslands could have encountered domestic horses and horse-drawn carriages earlier than the Central Plains.

For instance, in the southeastern part of Inner Mongolia, the Harqin Banner, within a thriving ancient culture known as the Lower Xiajiadian Culture dating back 4,300 to 3,600 years ago, archaeologists found horse bones from this period, which were mostly identified as domestic horses based on DNA analysis. These horses likely had a complex origin, suggesting frequent exchange activities with neighboring regions at the time. The Harqin Banner is located at the eastern edge of the Eurasian grassland zone, and crossing the Yanshan Mountains to the south leads into the Central Plains. Nomadic groups carrying domestic horses and carriage technology continuously expanded within the grasslands, estimated to have reached the Mongolian Plateau around 4,000 years ago.

The region of the Lower Xiajiadian Culture is likely to have had easier access to the domestic horses and horse-drawn carriage technology spreading from Central Asia along the grassland belt.

Therefore, we can reasonably answer the previous question: Who were the builders of the Shimao ancient city defending against? They were likely defending against the increasingly powerful nomadic groups on the grasslands, who used horse-drawn carriages and horses to conduct swift raids on areas controlled by the Shimao Culture.

In verifying this answer, genetic technology once again reveals its fascinating side. Didn't the Shimao ancient city yield skeletal remains of sacrificial victims? Scholars conducted DNA tests on 76 human skulls unearthed at the Shimao site and compared them with DNA from remains of ancient individuals in other regions at the time. The results indicated clear differences between these deceased individuals and the population of the inner city area of Shimao. This suggests that these victims were not "locals" of the Shimao ancient city. The genetic relationship of the victims in the sacrificial pit was closer to the population of the Lower Xiajiadian Culture, indicating that their homeland was in the grasslands to the north or northeast of Shimao.

The Shimao Culture population either had intense conflicts with the Lower Xiajiadian Culture population to the northeast or fierce conflicts with other nomadic groups on the northern grasslands. These nomadic groups had genetic similarities with the Lower Xiajiadian Culture population but were more distantly related to the Shimao Culture population. The unfortunate victims in the sacrificial pit were possibly war captives captured by the Shimao Culture population during conflicts or individuals seized from raids on the grasslands. Although modern genetic technology cannot alter their tragic fate, it can at least provide insight into their place of origin.

So, who were the "locals" of the Shimao ancient city? Genetic analysis suggests that they were closely related to the population of the Taosi Culture, which inhabited the central and southern parts of Shanxi province at that time. These populations likely emerged from the expanding and diffusing agricultural societies during the Yangshao Warm Period. The spread of domestic horses and carriage technology on the grasslands created a formidable opponent for the agricultural societies. The confrontation and integration between agricultural and nomadic societies throughout history persisted until modern times.

Whether belonging to the agricultural populace or the nomadic clans, both find their place within the biological realm of *Homo sapiens*, the modern human species. These lineages trace their ancestry back to the ingenious forebears of *Homo sapiens* who ventured forth from Africa. Their cognitive prowess and sagacity align harmoniously, placing them on a parallel tier of intellectual prowess. Hence, the mastery of equestrianism and the art of horse-drawn carriages does not exclusively grace the skilled hands of the nomadic cohorts traversing the grassy plains. The agrarian communities gracing the expanse of China's terrain have, since antiquity, upheld the tradition of pragmatism, promptly dismantling the nomads' dominion over the technological domain of domesticated horses and horse-drawn carriages.

Commencing from the twilight of the Shang Dynasty, the region of the Central Plains embarked upon an era characterized by domesticated horses and horse-drawn carriages. The different clans vying for supremacy on these plains increasingly enlisted cavalry and chariots in their pursuit, thereby elevating the frequency of employing domesticated horses and horse-drawn carriages in their daily lives. The agrarian societies' reverence for these equine creatures rivaled that of their nomadic counterparts.

Consider, for instance, a site nestled in the realm of Xianyang City, Shaanxi province, where the vestiges of four domesticated horses from the Late Western Zhou Period emerged. The findings of mtDNA analysis conducted on these four equines yielded a tapestry of fascinating revelations. These four steeds displayed distinct genetic profiles, with a pair among them bearing semblance to the genetic makeup of domesticated horses originating from the expanse of Central Asia and Europe. This revelation, of course, does not imply an odyssey of equestrian travel bridging Europe to the reaches of East Asia. Instead, it unveils a tableau of vibrant horse trade woven into the tapestry of yore. Horses from disparate domains frequently changed hands, and across successive generations, specific equine strains surfaced in distant realms situated thousands of kilometers apart, their genetic imprints accompanying them on this voyage. The variegated provenance of these four domesticated horses resonates with the rhythmic cadence of frequent horse trade between the Western Zhou Dynasty and the adjacent populations, intertwining seamlessly with other strands of trade.

The genetic chronicles of these horses further narrate an enchanting tale—each of the quartet possesses a resplendent chestnut coat, their countenances a symphony of uniformity. Evidently, the custodians of these steeds meticulously elected for a uniform chestnut hue. During the epoch of the Western Zhou, the upper echelons of nobility adhered to a corpus of exacting rituals. Within the tapestry of these ceremonials, an affinity for a specific equine coat color likely found its place, a preference that transcended genetic diversity. Thus, the genetic variances in the profiles of these four equines, and yet the consistency in their coat hues, might be attributed to this predilection.

By the juncture of the Shang and Zhou Dynasties at the latest, the vast expanse of China had gracefully transitioned into an epoch infused with the resonance of domesticated horses and elegantly adorned horse-drawn carriages.

4.3 Bronze: The Glory of Ancient Kings

The rocks on the Earth's surface undergo weathering, breaking down into soil, enabling plants to absorb nutrients and humans to cultivate farmland for sustenance. In fact, many rocks contain various metallic elements such as gold, silver, copper, iron, tin, lead, zinc, and more. These metals rarely exist in their pure form in nature; they often form compounds that are firmly embedded within the rocks.

To extract and refine these metals from rocks, humans have relied on the use of fire and metallurgical techniques. The era of humans using fire to create pottery dates back very far in history. They consciously experimented with using fire to treat various materials, including rocks with higher metal content. During the high-temperature roasting process, minerals like malachite could easily be oxidized and reduced by charcoal, melting them into metallic copper. This might be the origin of copper metallurgy.

From archaeological discoveries around the world, it can be seen that the earliest appearance of metallurgy occurred in the Western Asia region, where the earliest copper products were born. Foreign archaeologists have compiled data on the earliest discovered copper tools and their production dates in Western Asia, which were primarily concentrated in Turkey, Northern Syria, Northern Iraq, and Iran. Among them, the earliest copper artifacts in Western Asia can be traced back 9,000 years, when the region had already entered an agricultural society. At that time, people mainly used native copper, such as naturally occurring red copper drills, copper wire pins, malachite beads unearthed from the Çatalhöyük site in Turkey, and red copper needles from Sialk in central Iran.

Around 8,000 to 6,000 years ago, the metals observed in the Western Asia region were primarily red copper, with a few exceptions of lead, silver, and iron items. These were mostly small tools or ornaments, such as axes, drills, awls, beads, pieces, knives, shovels, buckles, tubes, chisels, rings, bracelets, seals, and scepter heads, and they all employed forging techniques. During this period, items related to casting started to appear in human settlements, such as crucibles, slag, ore, and smelting remnants.

Arsenic copper was one of the earliest binary alloy technologies mastered by humans, and its development also occurred in Western Asia. Naturally occurring arsenic copper minerals made smelting relatively easy. Starting around 6,000 years ago, arsenic copper emerged on the Anatolian Plateau of Western Asia and became popular in the surrounding regions, spreading metallurgical techniques even to the neighboring Central Asia, close to East Asia.

Moving forward from the smelting of red copper and arsenic copper, the next step was the smelting of bronze. Bronze is an alloy made by adding small amounts of lead and tin to copper, significantly enhancing the properties of copper products. Compared to red copper, bronze has a lower melting point, higher hardness, and is easier to cast into shapes. Bronze artifacts made from bronze have intricate and finely detailed surface decorations.

In ancient Babylonia over 5,000 years ago, the earliest bronze artifacts in the world appeared in the southern part of the Mesopotamian region. These mainly comprised tools and weapons made from copper-tin alloys. Around the same time, approximately 5,000 years ago, bronze artifacts also emerged in ancient Egypt, marking the beginning of the Bronze Age. Similar to the rapid dissemination of chariot and horse technologies mentioned earlier, bronze technology also spread quickly to surrounding regions.

The appearance of bronze artifacts on the land of China dates back very early as well. In a site located in the southern part of Gansu province, dating back around 5,000 years, a fully preserved bronze knife was unearthed. This 12.5 cm-long knife was crafted using the lost-wax casting method. This bronze knife is currently the earliest known bronze casting in China. Unfortunately, during that time, this was the only bronze artifact discovered in the land of China. Therefore, scholars suspect that this bronze knife might not have been locally manufactured.

One possibility is that due to frequent interactions among various groups across the Eurasian continent, this bronze knife was brought to the western part of China from Western Asia or Central Asia. Another possibility is that the local ancient population accidentally obtained high-quality tin-containing copper ore, which happened to be transformed into this bronze artifact during the smelting process. Regardless of the possibility, it indicates that around 5,000 years ago, the ancient people in China had not truly mastered the technology of bronze smelting.

The technological barriers will eventually be broken. After thousands of years of exploration and learning, the prosperous and distinctive Bronze Age has finally arrived on the vast land of China.

Around 4,000 years ago, with its core in the eastern part of Gansu and radiating to the northeastern part of Qinghai, Inner Mongolia, and Ningxia, there existed a powerful culture known as the Qijia Culture. Its metal smelting technology had matured, capable of producing intricately crafted bronze artifacts such as single or double-eared vertical-holed copper axes, button-adorned bronze mirrors, human-faced bronze utensils, ring-handled knives, and other complex bronze items. Particularly, they extended the use of bronze technology to create decorative elements for the human body and attire, expanding the applications of bronze artifacts. In the early stages of the Qijia Culture, red copper items were predominantly crafted. As time passed, their techniques continuously evolved. By the late period of the Qijia Culture, they were primarily engaged in producing tin bronze items. Of course, for the people of the Qijia Culture, bronze was still a precious commodity, so they might have been reluctant to use bronze for the production of high-consumption long-range weapons like bronze arrowheads.

Slightly later than the Qijia Culture, the Erlitou Culture further advanced in bronze technology. What's more critical is that the Erlitou Culture was situated in the Central Plains of China, an area rich in mineral resources. Once it mastered advanced metallurgy, the bronze culture it created surpassed the Qijia Culture and other contemporaneous cultures. The bronze culture of the Chinese land moved beyond mere imitation and entered a phase of active technological innovation, transforming the landscape of the Bronze Age.

The Erlitou site is located in the eastern part of the Luo River Basin in the middle reaches of the Yellow River, backed by the Kuang Mountain (邝山) and overlooking the Song Mountain to the south. The Yi and Luo Rivers flow to its south. Historically, this place was acclaimed as the "center of the world," a battleground for strategists and a site for imperial capitals. The Erlitou Culture emerged as a significant presence in this coveted area.

The most intriguing technological innovation of the Erlitou Culture is the "composite mold" technique. In other regions outside of ancient China at that time, most bronze artifacts were produced using either the forging method or the lost-wax casting method. The lost-wax casting involves crafting a wax model of the desired bronze object, covering it with clay inside and out, then baking it at high temperatures to melt and drain the wax, leaving a clay mold. Molten metal is then poured into the mold, and once cooled, the bronze artifact is obtained. The lost-wax method is relatively straightforward, suitable for producing simple-shaped bronze weapons, ornaments, and tools such as bronze knives, arrowheads, or plowshares. However, it's challenging to create large, intricate bronze artifacts using this method.

The composite mold technique is a complex bronze process where artisans create an inner mold and at least three outer molds from clay, assembling them to form a cavity. Molten copper is poured into the cavity and cooled to shape. This method can produce irregular and diverse complex objects, especially large bronze artifacts.

From the perspective of modern technology experts, creating a reverse cavity by enclosing multiple outer molds requires sharp spatial imagination. The dimensions, shapes, and decorations of the molds must be interrelated, and design must integrate with the casting process. This undoubtedly requires a standardized process system, achieving standardization, collaboration, and controllability in the production process. The Erlitou Culture had reached a level of technological and managerial sophistication akin to modern industry.

As a result, archaeologists excavated a wealth of bronze ritual objects at the Erlitou Culture site. Advanced techniques were applied to produce ceremonial items, such as bronze vessels used in drinking rituals. Bronze wine vessels appeared in the late period of the Erlitou Culture, initially starting with small-scale vessels resembling pottery jues, and later evolving to include warm wine vessels. This indicates that our ancestors might have had drinking rituals and a wine culture even before the introduction of bronze technology to the Central Plains. They likely initially used clay to create relevant items, and with the arrival of bronze technology, bronze, with its precious and superior material, replaced clay.

In addition to dings, the bronze ritual objects of the Erlitou Culture were all imitations of ceramic wine vessels, forming a group of bronze ritual items. However, bronze remained a precious commodity, and the lower classes had no access to it. Bronze ritual items were only found in the tombs of the upper-class nobility of the Erlitou Culture, serving as symbols of their social status and position.

During the Zhou Dynasty, various noble titles like duke (公), marquis (侯), count (伯), viscount (子), and baron (男) were established, and "jue" (爵), which originally referred to wine vessels, was used to indicate rank and status.

A small number of bronze weapons were also unearthed at the Erlitou site, including spears, battle-axes, hatchets, and arrowheads made of bronze material. However, among the melee weapons, only four specimens of spears, battle-axes, and hatchets were discovered, none of which had sharpened edges. Therefore, they were merely shaped into weapons, essentially resembling ceremonial items rather than practical weapons. It seems that until the time of the Erlitou Culture, bronze in the Central Plains was not extensively used for weapon production.

The Erlitou Culture, located in the Central Plains, and the Qijia Culture, situated in the northwest, overlap in their chronological periods. There was cultural exchange between these two cultures. Archaeological findings in Qijia Culture sites have yielded pottery and imitations with Erlitou Culture styles. The distinct "Turquoise-Inlaid Bronze Plaques" (镶嵌绿松石兽面铜牌) of the Erlitou Culture have also been discovered in the Gansu region during that time. While Qijia Culture artifacts have not been found in Erlitou Culture sites, there's reason to speculate that Qijia Culture might have influenced Erlitou Culture. The two cultures likely had metallurgical exchanges, with Qijia Culture, slightly preceding Erlitou Culture, potentially transmitting some bronze technology, thereby stimulating further innovation in the latter.

So, where did the bronze technology of the Qijia Culture originate?

Looking at the global context of that time, around 4,000 years ago, ancient civilizations worldwide were undergoing transformative changes. Whether it was the Nile civilization, Mesopotamian civilization, Minoan civilization in the Mediterranean, the Indus Valley civilization, or China's Xia civilization, urbanization was accelerating, writing systems were emerging and becoming more systematic, and metallurgical techniques were maturing. The development of metallurgy in the northwest region of China is undoubtedly tied to technological exchanges with Central Asia. The Qijia Culture was certainly influenced by the global civilization transformation, and its bronze technology likely came from the more western parts of the world.

As for the route through which bronze technology entered China, it might have been transmitted through the prehistoric Silk Road of Xinjiang, or it could have come from the grassland areas where the people of Shimao lived restlessly. Archaeological discoveries confirm that around 3,000 to 5,000 years ago, several early Bronze Age cultures were widely distributed in the Eastern Eurasian grassland belt. They shared common features such as early bronze smelting and production techniques represented by animal-patterned bronze ornaments and bronze short swords. This indicates that bronze technology spread rapidly across the grasslands, with various grassland communities mastering it. They likely disseminated this technology in their interactions with neighboring regions, even beyond the grassland areas.

From the bronze artifacts found in various regions, it's evident that the northern Chinese grassland belt encountered bronze technology earlier than the Central Plains. The Bronze Culture of the Central Plains was influenced by the grassland belt. For instance, in the Erlitou site, a long, narrow, thick-edged bronze weapon was unearthed. This type of weapon was previously unseen in the Central Plains and was actually a bronze battle-axe style popular on the northern grasslands. Similarly, the Erlitou site also yielded a ring-pommel dagger with engraved patterns on its hilt, which was also a weapon style from the northern grasslands. Tracing further west along the grassland belt, similar bronze battle-axes have been found in West Asian Iran, dating back several centuries, if not more, earlier than the Erlitou Culture.

Horses, horse-drawn chariots, and bronze weapons, before the advent of agrarian societies, greatly enhanced the combat capability of grassland communities. It's imaginable that when these grassland people drove chariots pulled by horses and wielded bronze battle-axes, the threat was substantial, prompting the ancient inhabitants of Shimao to construct more fortified cities to defend against external attacks.

Simultaneously, the agrarian cultures of the Central Plains and their surroundings should hold a degree of gratitude toward the grassland communities. It was the grassland communities along the grassland belt who brought these precious domesticated animals and advanced technologies to the east, leading to a profound transformation in the Central Plains Culture.

4.4 The Path Across the Grasslands: A Symphony of Fusion Where the Western Breeze Dances in the East

When it comes to ancient China's interactions with the outside world, we often think of the Silk Road and the remarkable journey of Zhang Qian to the Western Regions.

During the Western Han Dynasty, Emperor Wu of Han (Liu Che) dispatched Zhang Qian as an envoy to establish contacts with the Great Yuezhi (大月氏) in the Western Regions to counter the Xiongnu threat. Despite enduring numerous hardships on his journeys, Zhang Qian's efforts indirectly paved the way for communication between the Han Dynasty and the various states in the Western Regions, an achievement known as "piercing through the Western Regions" (凿空西域).

Through trade and interaction, ancient China established connections with Central Asia and even regions further to the west. Ancient merchants traveled through desert oases along the northern and southern foothills of the Tianshan and Kunlun Mountains, bridging the gap between the Eastern and Western worlds. These trade routes were later named the Silk Road, named after the exquisite Chinese craftsmanship of silk.

Due to the profound influence of the concept of the "Silk Road," scholars have long assumed that ancient China's trade and cultural exchanges with the West primarily followed the routes through the Tianshan and Kunlun Mountains in what is now the southern part of Xinjiang. China has revered jade since ancient times, with Kunlun jade, also known as Hetian jade (和田玉), from the northern foothills of the Kunlun Mountains being highly regarded. The phrase "Jade comes from Kunlun peaks" (玉出昆冈) embodies this tradition. The influence of Kunlun jade culture further reinforced the perspective that the East–West exchange route passed through the southern part of Xinjiang.

"The world had no roads, but many people walked, and thus, roads were made." Determining the path of early exchanges between the East and West is of extraordinary significance in understanding the cultural diffusion of the ancient world. The initial assumptions about the Silk Road led to some academic confusions, such as the origins and transmission of painted pottery.

The first group of scholars to systematically study cultural exchanges between ancient China and the Western world were Western scholars who came to China in the modern era for research, guidance, and teaching, such as the Swedish scientist Johan Gunnar Andersson (1874–1960). The first chapter of this book mentioned that the teeth of two Peking Man individuals unearthed from the Dragon Bone Hill site in Zhoukoudian were excavated by Andersson and others. However, Andersson's greater discovery during his time in China was the Yangshao Culture. Do you remember the "Yangshao Warm Period" mentioned in our previous chapter? This warm period spanning thousands of years marked the agricultural era on the land of China, and the Yangshao Culture thrived during this period.

In Yangshao Village, Henan province, Andersson unearthed relics of ancient China. The people of the Yangshao Culture were skilled in crafting exquisite pottery, adorning the surfaces of their pottery with vibrant and colorful geometric patterns or animal motifs using red and black pigments. Patterns such as human faces, fish, deer, frogs, and birds were the most common. Andersson and other Western scientists compared the painted pottery of the Yangshao Culture with that of ancient cultures in Central Asia and West Asia, suggesting that the painted pottery, represented by the Yangshao Culture, was introduced from foreign cultures. This advanced agricultural group, known for their Painted Pottery

Culture, might have migrated across mountains and rivers to the Yellow River Basin, where they blended their own local culture, resulting in a new prehistoric Chinese culture. They speculated that this group and their culture might have journeyed through Xinjiang and Gansu regions. Besides painted pottery, items like tripods, braziers, small-mouthed pointed-bottomed bottles, stone rings, and shell rings discovered from the Yangshao Culture might also have originated from the West.

The viewpoints of these Western scholars are referred to as "Sino-Babylonianism." During the first half of the twentieth century, when China faced invasions from foreign forces, some Chinese scholars sought to reflect on their own cultural issues and learn from the West, while many other scholars firmly believed in the excellence of their own culture, striving to uphold the thousands of years of continuous tradition. Amidst these mixed ideologies, the notion of Sino-Babylonianism sparked significant controversy among Chinese scholars. However, those who opposed Sino-Babylonianism lacked modern scientific knowledge and could only resort to citing ancient texts to counter, resulting in a somewhat awkward situation of "speaking past each other."

In the latter half of the twentieth century, indigenous archaeology emerged in China, allowing Chinese scholars to use modern science to discern the accuracy and distortions of foreign claims. They discovered that the appearance of Painted Pottery Culture on Chinese soil exhibited an interesting phenomenon: it emerged earlier in the east and later in the west. For instance, colored pottery in Gansu originated from the Guanzhong region of Shaanxi, and as one moves westward, the appearance of colored pottery becomes progressively later. Similarly, the colored pottery found in Xinjiang largely originates from Gansu and appears at a later time.

According to common sense, cultures tend to spread from earlier areas to later ones. Therefore, Andersson's viewpoint that painted pottery came from the west through Xinjiang and Gansu into the Central Plains doesn't hold up. Chinese scholars asserted that the painted pottery of the Yangshao Culture was independently created by our own ancestors, not brought from the West. This, in turn, supports the broader notion that Chinese civilization originated locally and was not imported from the West. This is referred to as the "Indigenous Chinese Civilization Theory."

Looking solely at the timeline of the appearance of painted pottery in China, the path depicted by the Indigenous Chinese Civilization Theory—spreading from east to west, from the Central Plains to Xinjiang—seems plausible. However, when analyzing the global distribution and chronology of ancient colored pottery, contradictions arise.

The distribution of ancient colored pottery worldwide extends from the East Asian Yellow River Basin to the west, stretching across Gansu, Qinghai, Xinjiang in China, through Central Asia, West Asia, and further to Eastern Europe and Southeastern Europe, forming a "Colored Pottery Cultural Belt" spanning over ten 10,000 kilometers. It's difficult to imagine that colored pottery in various regions was independently created without any exchange or borrowing taking place.

In terms of chronology, around 8,000 years ago, colored pottery first appeared in Western Asia, specifically in Iran and Iraq, followed by Syria, Turkey, China, Greece, Pakistan, Egypt, and other regions. If we were to map the appearance times of these colored pottery locations, a pattern emerges: the further away from Iran and Iraq, the later the appearance of colored pottery. Moreover, early colored pottery in many areas shares similar pigments and designs, suggesting that the origin of the global colored pottery culture lies in the region of Iran and Iraq, which then spread outward, eventually forming the Colored Pottery Cultural Belt. Of course, as Colored Pottery Culture spread to different parts of the world, it merged with local cultures, resulting in some localized styles and designs.

This leads to a contradiction. Following the trend of global colored pottery transmission, it would be expected that colored pottery traveled from West Asia through Central Asia to China, moving from west to east. According to this perspective, Xinjiang in the westernmost part of China should have been the earliest to have colored pottery, followed by Gansu and the Central Plains. However, the actual transmission of colored pottery within China occurred from east to west.

How can we explain this contradiction in the direction of colored pottery transmission?

The crux of the contradiction lies in an incorrect assumption about the input route.

China's colored pottery culture was indeed inspired by Western colored pottery, but it didn't originally arrive through Xinjiang and Gansu into the heartland of the Central Plains. Instead, it was transmitted through the "Grass Road," which ran from Northern Central Asia along the grasslands eastward, then southward from the Mongolian Plateau, crossing the Yin Mountains of Inner Mongolia to reach the Central Plains. Although this route might seem more circuitous compared to the direct route through Xinjiang and Gansu, in ancient times, the Grass Road was easier to traverse, resulting in a shorter journey from Central Asia to the Central Plains and lower costs.

The ancient Silk Road that passed through Southern Xinjiang was often referred to as the "Oasis Road" (绿洲之路) or "Front Silk Road," involving treacherous crossings of the Pamir Plateau. Eminent monks like Faxian in the Wei-Jin

Period and Xuanzang (Mōkṣadeva) in the Tang Dynasty chronicled their arduous journeys along the Silk Road, recounting near-death experiences. Travelers had to navigate vast desert perimeters, such as the Taklamakan Desert, trekking cautiously between widely spaced oases.

The "Grass Road," in contrast, was much more manageable. The terrain was gentle, without vast deserts or steep mountains to obstruct the path. A few relatively low mountain ranges needed to be traversed, and these ranges had multiple low-altitude passes, making the journey easier. The grasslands were also rich in plant and animal resources, facilitating provisions. Once large animals like horses, cattle, and camels were domesticated, travel along the Grass Road became even smoother. The journey from the grasslands to the Central Plains was not as challenging as modern people might imagine.

Chapter 2 of this book discusses our ancestral "Steppe Hunters" as they migrated eastward along the grasslands. This period coincided with a warm climate phase, and one piece of evidence is the changing water levels of the Juyan Sea in Western Inner Mongolia. The Juyan Sea is located to the north of the grasslands and adjacent to the Loess Plateau to the south, with abundant nearby water and grass resources. Furthermore, the second-largest inland river in China, the Etsin Gol, originates from the Qilian Mountains in Qinghai, flowing from south to north through Gansu and into Western Inner Mongolia, ultimately reaching the Juyan Sea.

The Etsin Gol, also called the "Heihe River," was known as the Black Water or Ruo Shui in ancient times. Unlike many major rivers in China such as the Yangtze, Yellow River, and Heilongjiang, which flow from west to east (latitude-wise), the Etsin Gol is a "meridional river," flowing along a north-south direction, connecting the northern edge of the Qinghai-Xizang Plateau, the Hexi Corridor in Gansu, and the grasslands of Inner Mongolia. The Etsin Gol River Basin has rich flora and fauna resources and has historically served as a crucial north-south passage in the northwest region of China, until environmental degradation led to the complete disappearance of the Juyan Sea.

By correcting the transmission route of the Painted Pottery Culture, the earlier mentioned contradictions are resolved. The Painted Pottery Culture centered around Iran and Iraq spread into the heartland of China through the grasslands and then expanded outward from the Central Plains to surrounding areas, including westward to Gansu and Xinjiang.

With a clear understanding of the transmission route of the Painted Pottery Culture, we can now revisit the westward diffusion of bronze technology. After the birth of copper smelting in West Asia, its spread was quite similar to that of the Painted Pottery Culture. Scholars have created maps of ancient world bronze technology distribution, which reveals a sequence of appearances in regions including ancient Babylonia in the Tigris-Euphrates Basin, ancient Egypt, ancient India, and finally ancient China. Clearly, the further the distance from the origin of copper smelting and bronze technology, the later these technologies appeared.

Looking at the distribution of Chinese bronze technology, the earliest bronze artifacts and smelting sites are mainly found in the northwest, the Central Plains, the northern grasslands, and the Eastern Shandong region (referred to as the Haidai Region 海岱地区 by archaeologists). In the early period of China's Bronze Age, the largest number of bronze artifacts was found in the northwest region, with thousands of artifacts unearthed. The Central Plains followed, with hundreds of bronze artifacts discovered in sites like Erlitou Culture. The northern grasslands, such as Inner Mongolia, and the Haidai Region had the fewest bronze artifacts. In summary, early Chinese bronze artifacts generally decreased from west to east.

However, there are signs of the transmission of bronze artifacts and technology from east to west in Western Regions. For example, in Gansu, the eastern bronze artifacts are dated earlier than those in the western parts, showing a phenomenon of bronze technology spreading from the Eastern to Western Gansu and even further to the eastern parts of Xinjiang. Therefore, scholars speculate that bronze technology first entered the northern grasslands, then spread southward to Eastern Gansu in China, and subsequently dispersed eastward and westward. The transmission of bronze technology is similar to that of the Painted Pottery Culture's entry and diffusion into China.

Ancient exchanges between the East and West were extensive, and using a term from today's computing, the introduction of Painted Pottery and bronze technology appears to have been a form of "resumable download," possibly occurring through multiple instances. For instance, after the birth of Painted Pottery technology around 8,000 years ago, it might have taken about 700 years for it to enter the Chinese mainland for the first time. However, the styles of Painted Pottery during that time and those that came later were notably different, yet both had connections with Western styles from different periods, suggesting ongoing exchange between East and West. Over thousands of years, the transmission of bronze technology from the grasslands to the south may have been even more frequent.

Another item that was brought into China from the grasslands that is particularly puzzling is tropical seashells.

At the Erlitou archaeological site, seashells have been unearthed, primarily found accompanying burials in the tombs of nobles. These shells are perforated, allowing them to be strung together with silk cords and worn around the neck and chest. The higher the specification of the tomb, the more seashells were used.

During that era, seashells were used as symbols of currency and wealth among the upper classes. This practice persisted at least until the Shang Dynasty, as evidenced in oracle bone inscriptions which recorded the award of seashells by the Shang king to certain nobles as funding and rewards for completing specific projects.

Initially, scholars believed that these seashells likely came from China's nearby seas, introduced to the Central Plains area through trade with the Dongyi (东夷) tribes along the eastern coastal region. However, the identification of seashell species revealed that these seashells belonged to warm-water varieties, with their origins in the tropical waters of the Indian Ocean and the South China Sea. They would never have lived along the northern coast of ancient China's East Sea.

So how did these seashells find their way to the Central Plains? After being harvested along the Indian Ocean coastline, they were traded northward into Central Asia and then relayed southward to the Central Plains region through the network of the Eurasian steppe. Seashells from the southern oceans took an unexpected journey, entering from the north. Later, as bronze coins gradually replaced seashells in the Central Plains region, the tradition of using seashells persisted among the nomadic tribes of the northern grasslands, such as the Xiongnu and Xianbei peoples.

Colored pottery, bronze, seashells ... the lives of ancient Chinese were enriched by these technologies and resources originating from the west and south. However, what the Silk Road truly delivered to the land of China might not have been these dazzling artifacts, but rather ...

4.5 Wheat: Rising to Prominence Over Time

Traditionally, the northern regions of China have favored wheat-based foods as their staple diet. Provinces like Shanxi, Henan, and Shaanxi each have their own distinctive wheat-based dishes. The primary ingredient for these dishes is flour made from wheat, making wheat the undisputed staple grain of the northern regions for over 2,000 years. As we learned in the previous chapter, in prehistoric times, the native crops of the land were foxtail millet from the north and rice from the south. Rice even made its way north during favorable climatic periods and was cultivated to some extent in the northern regions.

So, how did wheat rise to prominence later on, pushing back the northward movement of rice and dethroning the position of foxtail millet as the chief grain in the north?

Firstly, let's understand how wheat managed to come to the forefront.

The homeland of wheat lies in distant West Asia. West Asia, West Asia, West Asia ... Why is it that whenever we talk about a foreign technology or product, it often originates from there? The fundamental reason is that West Asia is the birthplace of human agriculture. The sparks of the Agricultural Revolution were ignited in West Asia, leading to successive waves of technological revolutions that spread globally.

The earliest discovered remains of wheat date back to around 10,000 years ago. The initial domestication of wheat involved two varieties. Initially, an ancestor of wheat was tamed, which then hybridized with a type of goat grass, resulting in a second variety. However, neither of these is the wheat we consume today. Around 8,000 years ago, this second variety of wheat spread eastwards to the Iranian Plateau. There, it hybridized with local wild grasses, leading to the emergence of a new cultivated wheat species known as hexaploid wheat. This is the wheat variety widely cultivated by people around the world today, covering 95% of wheat-growing areas. It's often referred to as common wheat or bread wheat. "Hexaploid" refers to the fact that this wheat's genome consists of three sets (six copies) of similar, yet distinct genomes integrated together. In contrast, the human genome is considered "diploid," consisting of genome sets inherited from both parents.

The appearance of hexaploid wheat can be seen as a significant "genetic engineering" event in ancient times. This wheat was both delicious and high-yielding, a treasure that quickly gained popularity. Several ancient civilizations, such as the Mesopotamian civilization in the Tigris and Euphrates Valleys, ancient Egypt along the Nile, ancient India along the Indus River, as well as later civilizations like ancient Greece and Rome, all built their prosperity upon agriculture centered around wheat as their primary grain. Until the Industrial Revolution, wheat remained the world's primary staple crop.

Even as early as 7,000 years ago, wheat had already made its way eastward to Central Asia. For example, in the foothills of the Köpet Dag mountains in what is now Turkmenistan, wheat was being cultivated. However, after this initial progress, the spread of wheat to the east unexpectedly stalled, and this hiatus lasted for thousands of years. The shadow of wheat failed to cast itself over East Asia for a very long time. What was the reason behind this? Was it geographical barriers preventing wheat from reaching the Chinese mainland? Or did our ancestors simply prefer millets and rice, rejecting wheat?

The fundamental reason lies in the botanical traits of wheat that posed challenges for East Asian farmers.

Wheat's origin is in West Asia, in the Mediterranean climate zone. Readers familiar with basic geography know that the Mediterranean climate is quite peculiar. While most climates have cold, dry winters and warm, wet summers, the Mediterranean climate's rainfall pattern is the opposite: cold, rainy winters and hot, dry summers. Rainfall mainly occurs during the winter and spring seasons. This means that in this region, wheat is sown in winter and harvested in summer.

China's northern regions mainly experience temperate monsoon climates, while the southern regions have subtropical monsoon climates. The defining feature is simultaneous rain and heat during the summer, with hot, rainy summers and cold, dry winters. Precipitation is concentrated in the summer. But during the crucial spring season when wheat needs water the most, China experiences a dry period. This isn't favorable for wheat growth. And during the summer when wheat is ready to be harvested, excessive rainfall is detrimental to wheat maturation and harvesting.

The mismatch between wheat's growth cycle and China's indigenous climate hindered wheat's eastward progress, preventing it from entering the Chinese mainland for a long time.

Wheat, after all, can be considered a dryland crop, making it more suitable for the arid environment of northern China compared to rice. As for the issue of water shortage in the spring, it can be resolved through manual watering at the beginning of the season. After resting in Central Asia for nearly 3,000 years, wheat finally took its steps toward East Asia. The introduction of wheat led our ancestors to initiate a history of native irrigation agriculture and even required the construction of large-scale water facilities to meet the wheat's water needs.

When did wheat arrive in China? We can determine the timing through archaeological discoveries and historical records, but first, we need to focus on the Altai Mountains region in China.

Due to the proximity of the Altai Mountains to Central Asia, the cultivation of wheat and naked barley (highland barley) there began very early. Archaeologists found charred plant seeds in the Tongtian Cave site in the Altai Mountains and, upon analysis, identified both wheat and naked barley seeds dating back 5,200 years ago. Because the Altai Mountains are at higher latitudes than much of Central Asia, they experience shorter sunlight duration and lower temperatures, requiring species adaptation for wheat cultivation. Biologists speculate that after wheat was introduced to this area, it hybridized with local wild barley, obtaining crucial genes for thriving growth in high-latitude, high-altitude regions. This process allowed wheat to be domestically cultivated and become an essential crop in the region.

In other words, if we consider today's borders of China, wheat was introduced to northern Xinjiang over 5,000 years ago. So, when did wheat cross the deserts and mountains of the northwest to reach the Central Plains and its surroundings?

In the oracle bone inscriptions from the Shang and Zhou Dynasties, the characters for "来" (lai for pinyin; come in English) and "麦" (mai for pinyin; wheat in English) are already present, likely referring to wheat and barley, respectively. During that time, ancient people might have used the former to indicate wheat and the latter to refer to barley. In the "Zuo Commentary" (左传), there's a passage mentioning, "Zhou's older brother lacked wisdom, unable to distinguish beans from wheat, hence he cannot be appointed ruler." This narrative from the Spring and Autumn Period recounts the story of a noble in the state of Jin who was considered unfit to be ruler due to his inability to differentiate between beans and wheat (implying a lack of common knowledge). This indicates that wheat was already an important crop in the northern state of Jin during that time and was known to the general populace.

Analyzing these written records, it's clear that wheat had already been introduced and familiar to people in China over 3,000 years ago. Therefore, the introduction of wheat dates back even further into antiquity, requiring archaeological findings to provide more precise information.

Currently, the earliest discovered wheat remains in northern China come from Jiaozhou in Shandong province. Only one grain of wheat was unearthed, and radiocarbon dating places it at around 4,450 years ago. However, Jiaozhou lies southeast of the Shandong Peninsula, bordering the East China Sea. Geographically, it is more distant from the Central Plains and the northwest regions compared to the closer proximity to Central Asia.

As a result, scholars are still uncertain why the earliest wheat remains have been found in the coastal area of Shandong rather than the Central Plains or the northwest. It's possible that during that era, wheat was cultivated in various northern regions of China but in limited quantities, making it challenging for remains to persist in those areas. Jiaozhou might have simply preserved some remnants. Wheat, being a plant material, is prone to decay and is difficult to preserve over long periods at archaeological sites. Alternatively, there could be issues with the dating and information associated with that particular wheat sample.

Considering the timeline, it's reasonable to believe that wheat was introduced to the eastern part of the Eurasian steppe more than 5,000 years ago, serving as a supplementary food source for the pastoral lifestyle of the region. This timeline would precede its introduction to the Central Plains in the south. Therefore, we have reason to think that the Eurasian steppe served as a significant intermediary point on wheat's journey to the heart of China. Wheat likely entered through Central Asia into the northern Altai Mountains, subsequently spread to the eastern parts of the Eurasian steppe, and eventually traveled south to the border region between the Central Plains and the steppe, including the Yinshan and Yanshan mountain ranges. Finally, it crossed these mountains to reach the Central Plains.

Setting aside the enigma of Jiaozhou's wheat, around 4,000 years ago or perhaps a touch later, vestiges of wheat emerged in both the heartland of China and the northwestern reaches. Among these, wheat retrieved from several archaeological sites in Qinghai, Gansu, and the eastern expanse of Xinjiang bears witness to an age nearing 4,000 years past. Thus, within the narrative of wheat's eastern journey, the plains indeed stand as a pivotal station, while Xinjiang might well constitute an additional node along the trajectory of wheat's ingress.

As a cultivar, wheat's propagation found more fertile ground within agrarian communities. Contrasted with the allure that steppe societies afforded to domesticated horses, horse-drawn carriages, and bronze technology, the denizens of Southern Xinjiang bestowed their favor upon wheat, fostering verdant oases where every verdure-filled enclave burgeoned into an agricultural haven, harboring aspirations for high-yield crops. Emerging from Central Asia, should wheat surmount the Pamir Plateau's threshold, it would then traverse into the Tarim Basin. Although an enclave of desert, the basin's bosom harbors an array of oases along its north-south axis. Wheat, akin to an itinerant pilgrim, charted its course eastward through these verdant outposts, meandering into Gansu's fertile Hexi Corridor, a realm primed for cultivation, and further easterly to the heart of the Central Plains. This very itinerary would, in due course, etch its name upon the annals as the illustrious Silk Road.

With the passage over mountains and rivers into the cradle of the Central Plains, wheat found itself entwined with the legacy of our agrarian forebears and the vast expanses of arable terrains. Swiftly, an era marked by the dominion of wheat graced the vast expanse of China's heartland. The grand theatrical spectacle of wheat's ascension to a position of prominence unfolded, most notably upon the stage of the Guanzhong Plain, an expanse celebrated for its "Eight Hundred Miles of Qin Fields."

The Wei River, an esteemed tributary of the Yellow River, courses through the interstice betwixt the Qin Mountains and the Loess Plateau, meandering eastward to merge with the Yellow River. The siltation wrought by the Wei River bequeathed unto the landscape the fecund Guanzhong Plain. This demesne basks within a temperate belt, graced with a semi-humid climate and malleable loam. The generous flow of the Wei River and its tributaries conferred an irrigation lifeline. The Guanzhong Plain unfurled as an eminent agricultural dominion within Northern China.

In the throes of the Yangshao Culture, dating back approximately 6,000 years, the denizens of the Guanzhong Plain forayed beyond the mantle of a hunting-gathering subsistence, embarking upon the realm of dryland husbandry. During the nascent era of Yangshao, the primary cereals were millets—the foxtail millet and proso millet. Astonishingly, even rice—a crop native to the south—commanded a presence upon this terrain. Notable is the discovery of rice grains within the annals of an archaeological site in Weinan, Shaanxi, constituting a quarter of all grains unearthed, attesting to the prominence of rice cultivation. The warm and humid climate of the Yangshao epoch stands revealed as the unsung harbinger for rice's northern migration into the bosom of the Guanzhong Plain.

However, at the juncture of the Erlitou Culture, over 3,700 years ago, though wheat had taken root, the grains exhumed accounted for less than a percentile within the corpus of cereal remnants, rendering them a mere culinary interlude in the lives of our distant forefathers. As centuries unfolded, in the fissures of Central Plains' ruins, the percentage of wheat grains burgeoned to occupy 10% to 20% of the grain assemblage—a marked ascent positioning wheat as a contender, third in rank after millets. By the Western Zhou epoch, wheat grains unearthed in northern realms overshadowed other crops in number, solidifying its rank as the second preeminent crop. The Eastern Zhou epoch witnessed the emergence of wheat as the crown jewel in select realms, surpassing indigenous millet cultivation. Case in point lies in the Wei River Basin previously mentioned; here, the abundance of unearthed wheat grains surpassed an astonishing 80% of the total grain inventory.

In the wake of wheat's territorial conquests, rice gradually receded from the Guanzhong Plain. The cultivation of rice demands a copious water supply. Its initial foray into northern domains rested upon its superior yield in comparison to millets. Presently, wheat's yield stands poised for comparison with rice, coupled with a diminished thirst for water. This accords wheat an aptitude for broadscale cultivation in the north, thereby engendering a gradual retreat of rice to southern confines.

Ultimately, during the epochs of the Qin and Han Dynasties, wheat seated itself atop the pantheon of northern cereal crops. The age-old grain landscape of China transitioned from a "south of rice and north of millet" archetype to one characterized by "south of rice and north of wheat," an enduring legacy that persists to the present day.

4.6 Sheep and Cattle: We Are Here for Sacrifice

On the eve of China becoming an ancient civilization, two significant animals arrived in the land of China alongside wheat: sheep and cattle.

Domesticated sheep first appeared in West Asia, yes, West Asia! People might wonder if 10,000 years ago extraterrestrial beings descended upon West Asia, imparting ancient human civilizations with extraterrestrial technology for mining and animal domestication. However, it wasn't extraterrestrials who concocted this science fiction-like scenario, but rather the Agricultural Revolution.

Around 10,000 years ago, right as agriculture was taking its first steps, domesticated sheep emerged in the region of West Asia. Through genetic analyses of domestic sheep from various parts of the world, it's possible that the evolution of domestic sheep involved domestication events occurring more than once. In other words, after domestication occurred in West Asia, there were several instances of genetic exchange between domesticated and wild varieties, continuing the process of human-driven domestication. Among these domestic sheep genotypes, one holds a dominant position among East Asian sheep, gradually decreasing in frequency from East to West. Therefore, scientists speculate that domestication of sheep might have also happened in East Asia. This East Asian domestic sheep genotype is distinct from the one in West Asia.

The earliest discovered domestic sheep remains in China are likely from the previously mentioned Tongtian Cave site in the Altai Mountains. There, a considerable number of sheep bones were unearthed, alongside bones from small mammals and birds. Considering the discovery of wheat grains, it's conjectured that around 5,200 years ago, the early inhabitants of the grasslands engaged in mixed livelihoods—combining agriculture, animal husbandry, hunting, and gathering—to adapt to the natural environment of the grasslands and mountains.

Subsequently, domestic sheep remains were found in sites in the eastern part of Gansu and Qinghai, dating back around 5,000 years ago. By 4,000 years ago, domestic sheep bones were unearthed from multiple sites in the Central Plains region. Therefore, the spread of domestic sheep in China likely occurred from west to east. Around 5,000 years ago or even earlier, ancient people brought domestic sheep from West Asia into China, where these foreign domestic sheep interbred with local wild sheep. After domestication, this gave rise to the East Asian domestic sheep type.

By 4,000 years ago, the majority of domestic sheep in the land of China already belonged to the East Asian domestic sheep type, with only a small number retaining the West Asian type. Archaeological findings of domestic sheep remains in places like Gansu and Xinjiang suggest that West Asian domestic sheep, akin to colored pottery and bronze technology, were transmitted from the Eurasian steppe belt.

Having delved into the realm of sheep, let us now delve into the discourse of cattle. Did the earliest domesticated strains of the illustrious yellow cattle indeed trace their origins to the lands of Western Asia? This query, I must affirm, finds itself resting upon the precipice of partial accuracy.

Within the taxonomy of our forebears' domesticated bovines, the epithet "yellow cattle" encompasses all bovines excepting yaks and water buffalo, encompassing notably both the common and humpback breeds. The domestication of the common breed undoubtedly germinated in the heartlands of Western Asia, a historical juncture dating approximately 11,000 years in the past, thus revealing an early emergence. In contrast, the humpback lineage, deemed the progeny of the Indian subcontinent's Indus River Valley, came into the fold two millennia thereafter, lingering in temporal contrast to the common breed's domestication. By an incidental observation, the yak, an emblematic "ark of the plateau," met its domestication amidst the domains of Central Asia, its chronicle trailing just behind the common cattle.

In more contemporary epochs, the erudition bestowed by Chinese scholars yields revelations of bovine bone remnants unearthed in the province of Heilongjiang, whispers from the antiquity of ten millennia. The bones, adorned with telltale grooves upon teeth and mandibles, allude to the imprint of domestication. While the validation of this revelation remains pending, a tantalizing prospect emerges—that the northeastern precincts of China might stake a claim as the cradle of yellow cattle, its chronicle in no way dilatory to the expanse of Western Asia. Alas, a veil of uncertainty shrouds these grooves, which may bear varied explanations beyond the sway of domestication. In considering that Western Asia stands as the veritable birthplace of agriculture, replete with burgeoning populace and copious harvests, armed with the requisites for managing and nurturing sizable fauna, contemporary scholars uphold the assertion that "Western Asia preempts all as the cradle of early yellow cattle domestication."

Casting aside the nebulous genesis, an ancient tapestry dating 4,500 years prior beckons us to the Yellow River Basin. Within this tableau, the footsteps of domesticated yellow cattle imprint, a pageant that seemingly unfolds subsequent to the ovine entry. The archival keepsakes of bones from these domesticated yellow cattle are chiefly scattered across the Eastern recesses of Gansu, the northern and central reaches of Henan. Thus, the course of this bovine journey into China appears akin to the ovine pilgrimage—both traversing the verdant tapestry of the Eurasian steppe. After all, an abundance of emerald pastureways punctuated their sojourn, providing sustenance to these gentle beasts, while the sages of the steppe lent their sagacious stewardship. It's postulated that this first cohort of domestically raised yellow cattle upon the Chinese mainland largely foraged upon plants of the carbonaceous ilk, with a predilection for the likes of foxtail millet and proso millet—staple grains in the northerly climes. One might imagine the pastoral ancients, ingeniously employing the harvested stems and leaves as sustenance for their yellow cattle.

Molecular biologists, in their quest to uncover ancient truths, ventured to regions such as Xinjiang, Qinghai, and Inner Mongolia. There, they dissected the genetic relics within ancient bovine bones, a palimpsest of millennia past. The findings unveiled a panorama wherein all samples align with the common breed, standing in contrast to their humpback kin. This quirk of genealogy hints at a later ingress of humpback cattle from ancient India. Moreover, the genetic signatures predominantly herald from Western Asia, with scant whispers of northeastern provenance. This echoes the echoed tale of the sheep, whose lineage underwent its own veritable domestication within the tapestries of East Asia.

Contrary to the path trodden by domesticated sheep, yellow cattle also charted a course through the domains of the southwest. The tale begins with the taming of humpback cattle within the confines of the Indus River Valley, disseminating throughout South and Southeast Asia, before embarking upon the path to China from the southern province of Yunnan. A picturesque isle, christened Yinsuo Island, nestled within the ethereal embrace of the Erhai Lake in Dali, Yunnan, bore witness to the ancient echoes of yellow cattle, their remains interred within its hallowed soil.

As sheep and cattle waltzed onto the stage of the Chinese landscape, they bequeathed not only nourishment of flesh, dairy, and pelts to our forebears but also bestowed an intricate dance upon the tableau of their cultural mores.

It is etched within our cognizance that the porcine realm was indigenously cultivated upon the sacred soil of China, a partnership that unfurled some nine millennia bygone. Across the undulating contours of the Chinese landscape, echoes resonate from antiquity, revealing that ancestral ceremonies and rites were often accompanied by the solemn sacrifice of pigs—an occasional nod extended to the loyal canine companions. In a parallel melody, a symphony commences around four millennia ago, as the annals of antiquity unveil the emergence of sacrificial practices enriched by the offerings of cattle and sheep, weaving their significance ever more profoundly into the fabric of tradition. Thus, the historical ledger narrates a tale wherein these envoys from distant lands, bearing the nomenclature of cattle and sheep, etched their storylines upon the very parchment of culture, forever altering the cadence of custom and reverence.

With the advent of the Shang Dynasty, this practice of sacred offerings spiraled into an intricate art form, one that harbored intriguing tales of "animal discrimination." The annals of the Shang Dynasty lay witness to two grandiose ceremonies of sacrificial splendor—the "Tailao" (太牢), a regal ceremony befitting a monarch, necessitating a triumvirate of bovine, ovine, and porcine; and the "Shaolao" (少牢), a lesser rite catering to the aristocracy, obliging solely the company of sheep and swine. Thus, if we were to scale the hierarchy of reverence assigned to various creatures within the pantheon of ancestral minds, we'd encounter a roster where cattle, sheep, pigs, and dogs ascend the tiers in resolute sequence, with the yellow cattle reigning as the apical envoy.

The etymology of the term "sacrifice" (牺牲) serves as a vestige, inscribed upon it the indelible icon of "cattle" (牜/牛), a symbolic testament to the paramountcy of bovine offerings. Within the theater of divination, a practice revered during the Shang Dynasty, the scapula of yellow cattle found a newfound purpose as divinatory instruments. In the annals of artisanal craftsmanship, the bones of yellow cattle played a pivotal role, crafting the implements of bone within their bony grasp. The chime of history carried onward, resounding into the epoch of the Spring and Autumn Period and the Warring States Era. Here, as iron-forged implements unfolded their dominion, the robust and docile nature of yellow cattle found their call to service in the fields, ennobling them as steadfast allies of the toiling agrarian populace.

4.7 The Two Sides of Civilization Coins

By now, in the Central Plains region 5,000 to 4,000 years ago, the so-called "five grains" of ancient China—foxtail millet, proso millet, beans, wheat, and rice—were all abundant. The so-called "six domestic animals"—dogs, pigs, cows, and sheep—had emerged, while horses were galloping on the northern grasslands, and chickens might have been crowing at dawn in the southern regions; later, these two would also enter the Central Plains.

Among these crucial species, wheat, cows, sheep, and horses all came from the Western world. The immensely significant technologies of bronze and chariots also originated from the Western world, and similarly, the essential technique of colored pottery also came from the Western world. So, should we pick up the Sino-Babylonianism theory proposed by Johan Gunnar Andersson and others, dust it off, and present it on the academic table?

Logically and factually, we cannot so decisively support the Sino-Babylonianism. As discussed in the previous chapter, the northern ancestors of the Chinese tamed foxtail millet and proso millet, and around 7,000 to 5,000 years ago, they spread westward along the grasslands, allowing ancient people in western Asia and Europe to taste the flavors of millets. Dogs and pigs, both domesticated animals, originated in China and also spread to other regions of the Eurasian continent, including the Western world, at an early stage.

If we label Chinese civilization as "coming from the West" based solely on the introduction of certain species and technologies, then logically, considering the historical fact that ancient China exported species to other regions, we could also argue that Western civilization "came from the East."

So, is the theory of indigenous origin correct? Was our civilization born and thrived locally, and would Chinese civilization have flourished magnificently even without the influence of foreign cultures (including resources and technologies)?

The theory of indigenous origin goes to another extreme, as it fails to acknowledge the principle of the Tragedy of Tasmania. It assumes that a closed-off landmass of China could independently give rise to a splendid civilization. In reality, since the domestication of wild plants occurred in West Asia and East Asia, and the development of primitive agriculture began, East Asia ceased to be a large-scale isolated entity. Even in the earlier era of expansion of modern humans, East Asia was no longer an isolated island in terms of population and cultural exchange; there was constant interaction with the Eurasian continent. People moved in and out, carrying with them thoughts, cultures, technologies, and species.

Those who adhere to the theories of foreign origin or indigenous origin only see, or are willing to see, one side of the coin, neglecting or avoiding the other side.

Archaeologists and historians have their standards for defining a civilization. Different scholars might have slightly different criteria, but they generally include a few elements: metallurgy (bronze), urbanization, writing systems, and belief (religious) systems. An ancient civilization recognized by scholars might not possess all these elements, but it would have most of them.

For example, the Liangzhu Culture we discussed earlier already had urbanization and a belief system but lacked metallurgy and a writing system. Thus, mainstream foreign scholars did not elevate Liangzhu Culture to the status of a "Liangzhu Civilization." However, domestic scholars tend to believe that with the large-scale societal structure and governing capacity of the Liangzhu Culture, it deserves the title of "Liangzhu Civilization." Hence, whether the Liangzhu society qualifies as an ancient civilization or not remains a topic of debate.

From the perspective of civilization's standards and its development, the birth and establishment of Chinese civilization in East Asia were made possible by the abundant local population, food production, and cultural traditions of the Chinese land. It absorbed technologies, species, and even some ideas and cultures from the Western world, transforming itself and melting various regional cultures and technologies into the holistic Chinese civilization, stepping toward the "earliest China."

Bronze and domesticated horses were the decisive driving forces behind the birth of Chinese civilization as well as other civilizations. The bronze weapons and horse-drawn chariots derived from them provided ancient powers with long-range military transportation capabilities, which were crucial for the establishment of national political authority.

The transmission of bronze and chariot technologies to the northern steppe regions was primarily employed for military purposes, enhancing the combat capabilities of the steppe populations and exerting significant pressure on the southern agrarian societies. The defensive capabilities demonstrated by the ancient city of Shimao reflected this pressure. Later on, the diffusion of bronze and domesticated horses into the northern regions undoubtedly also boosted the combat strength of various populations in the Central Plains. In order to collectively defend against enemies, originally dispersed tribes would form larger and more powerful tribal alliances, breaking down ethnic and cultural boundaries.

Within these tribal alliances, bronze and chariot technologies introduced long-range military transportation capabilities. In the past, if a remote tribe initiated a rebellion, the alliance leaders had limited reach and struggled to suppress it. Now, they could swiftly dispatch an army equipped with bronze weapons and centered around horse-drawn chariots to quell the rebellion, deter wavering tribes, thus enhancing the internal cohesion of the tribal alliance, and progressing toward a civilized nation.

Furthermore, in the northern regions, bronze technology played a role akin to a catalyst for civilization. Bronze technology was a high-tech innovation on the eve of the birth of ancient civilizations. Our ancestors replaced old pottery with bronze vessels for rituals and ceremonies, creating the unique ritual and music culture of Chinese civilization. This elevated the cultural construction within tribal alliances, enhancing internal cohesion peacefully.

Simultaneously, the introduction of foreign species led to a greater demand for trade among various tribes in the northern regions, fostering close connections between them, a trend reflected in genetic studies of ancient livestock.

The introduction of superior breeds of cattle from the Western world into China, such as the high-quality yellow cattle, provided ancient populations with excellent labor and delicious food sources. Analysis of the genes of yellow cattle raised by various cultural groups of that time reveals the frequency of their trade interactions, indicating that previously distinct populations were becoming increasingly interconnected. In the northwestern, northeastern, and central regions at that time, the predominant yellow cattle genotypes were the same across a wide range of sites. A few other yellow cattle genotypes had lower frequencies, and their distribution varied among sites. This suggests extensive exchange of yellow cattle in the northern regions, where tribes eagerly adopted superior breeds of domestic cattle.

Interestingly, even ancient Chinese legends told stories of cattle and sheep trade. According to records in ancient texts like the "Bamboo Annals" (《竹书纪年》) and the "Classic of Mountains and Seas" (《山海经》), a king of the Shang Dynasty had an ancestor named Wang Gai, who along with his brother, drove cattle and sheep to the tribe of Youyi for business, only to be robbed and killed by the nobles of Youyi. Wang Gai's son, Shang Jia Wei, sought revenge but lacked the power. After four years, he borrowed the soldiers of the River God and finally wiped out the Youyi tribe, avenging his father's death. In the oracle bone inscriptions of the Shang Dynasty that followed, there were records of rituals dedicated to Shang Jia Wei, reporting the current affairs of the kingdom. Shang Jia Wei was clearly respected by the people of the Shang Dynasty, and his descendants often honored him with rituals.

While the details of ancient legends might not be entirely accurate, the scenario depicted in this legend of the Shang tribe engaging in long-distance livestock trade likely reflects the actual circumstances of intertribal interactions in the northern regions at that time.

This legend also reflects the turbulent state of the northern regions at that time, where peaceful commercial exchanges and violent clashes coexisted among various tribes. Scholars can deduce this tumultuous history from archaeological sites.

The Yangshao Warm Period was originally a joyful time for our ancestors. The weather was favorable, crops were abundant, new technologies and species emerged constantly, and people in the vast world spread their cultures to one another. Life was getting better and better. However, about 4,500 years ago, the northern regions entered the Longshan Period, a period lasting for several centuries, during which the harmony of the Yangshao Period collapsed.

From the remains of the Longshan Period settlements, we can see that communities began constructing sturdy moats and even larger settlements built defensive walls and structures to protect against neighboring threats and unite the tribe internally. The northern regions entered an era of numerous independent states constantly warring against each other.

What caused this transformation from the Yangshao Period to the Longshan Period? Clearly, population growth played a significant role. By the time of the Longshan Period, possibly millions of people lived in the heartland of agriculture, the Central Plains region. Given the productivity of the era's farming techniques, the yields from cultivated fields were not high. Crops like foxtail millet, proso millet, and beans were well domesticated and planted, while wheat was still on its way. The increase in population brought about intense survival pressures. Tribes strove to expand their living space, leading to conflicts over farmland, pastures, and water sources.

The climate of the Longshan Period was also one of the factors contributing to societal upheaval. After experiencing the warmth of the Yangshao Period, the Longshan Period found itself at the end of the Holocene climatic optimum, marked by climate fluctuations and occasional cooling events. These climate fluctuations triggered various natural disasters, sparking survival crises among certain tribes and driving them to violence and resource competition.

Ultimately, climate change acted as the "last straw" that propelled the ancient Chinese lands from a state of turmoil into the earliest stages of China.

Around 4,000 years ago, both the Chinese lands and the Eurasian continent seemed to be undergoing significant climate shifts.

For instance, the Liangzhu Culture, more than 4,000 years ago, entered a decline. Up the Yangtze River, the Shijiahe Culture and the Baodun Culture in the middle and upper reaches faced declines during the same period. The Qi Family Culture in the north disappeared around 3,700 years ago, and cultures in the Liao River Basin to the north also declined around the same time. Only the Erlitou Culture in the Central Plains region managed to rise amidst this widespread cultural decline.

Archaeologists refer to this dramatic transformation as the shift from a sky full of stars to a sparse sky with only the moon shining, symbolizing that various ancient cultures across the east, west, south, and north dimmed while a cultural moon rose over the Central Plains.

In my previous work, "Grass and Cereals—A 4,000-Year History of Chinese Civilization" (《草与禾—中华文明 4,000 年融合史》), I analyzed the climatic reasons behind the vast cultural changes, suggesting that the north-south movement of the West Pacific subtropical high-pressure system could have led to the extensive decline of ancient Chinese cultures.

If we broaden our perspective to the global scale, it appears that the entire Eurasian continent was undergoing significant climate changes during that time.

Around 4,200 years ago, the Ancient Egyptian Kingdom reached its end amidst the turmoil of a massive slave uprising. However, scholars have recently proposed a new perspective on the downfall of the kingdom. They examined coffin boards from that era and analyzed the tree ring patterns, discovering extreme and brief periods of drought during that time. The resulting crop reduction caused significant trouble for the rulers, eventually spiraling out of control.

In the Mesopotamian region, archaeologists excavated a disappeared large city and found evidence of severe drought in soil layers dating from 4,200 to 3,900 years ago. Even earthworms died in large numbers due to the drought.

Around 3,700 years ago, the Indus Valley civilization in India suddenly disappeared. While scholars previously thought that the civilization might have been invaded by Aryan people from the northwest, climatologists found evidence of a severe decline in the Indus Valley civilization during that time, likely due to unstable irrigation caused by climate fluctuations.

Why did the global climate experience such immense changes around 4,000 years ago? Currently, there is no definitive answer. However, what we can ascertain is that ancient civilizations such as ancient Egypt, Ancient Babylon, and ancient India, which had entered the stage of advanced civilization before ancient China, suffered heavy blows due to climate changes. In contrast, the Chinese lands managed to rise from the climate-induced challenges, and the Central Plains region entered the stage of advanced civilization.

4.8 The Earliest China, Ancestral Homeland

In the midst of tumultuous times, within the relatively habitable Central Plains region, the Erlitou Culture finally emerged, expanding its influence and ascending to the heights of civilization.

Archaeological findings reveal that cooling temperatures, floods, droughts, and other disasters during the late Longshan Period led to a decline in population size in the northern regions. Simultaneously, the population concentrated in the core area of the Central Plains, and around the central city of Erlitou, the population defied the trend and expanded. The excavated remains at the Erlitou site display a grand cityscape with intersecting thoroughfares, square and orderly palace complexes, along with various arranged structures within. This layout, characterized by a central axis, was inherited by successive ancient Chinese dynasties, such as the Han and Tang Dynasties in Chang'an, and the Ming and Qing Dynasties in Beijing.

The society of Erlitou during that time had already established a structured hierarchy. Nobles resided in the eastern part of the palace district, while the western and northern areas of the "city" were occupied by smaller ground-level or semi-subterranean buildings, serving as the living quarters for commoners. Within the city, numerous functional zones existed, including a bronze casting workshop. The bronze workshop at Erlitou was sizable and operated for an extended period, featuring casting workshops and kilns for firing pottery molds, indicating a high level of specialization in bronze casting techniques. Between the palace district and the bronze workshops, there was also a workshop for producing turquoise artifacts. The ancestors of Erlitou primarily crafted decorative items like turquoise tubes, beads, and inlays there.

In the central and eastern parts of the city, various structures and relics related to religious and ceremonial activities were concentrated. These included circular ground structures, rectangular semi-subterranean buildings, and associated tombs. It's possible that the ancestors of Erlitou performed ancestral and deity worship at these ceremonial sites. They might have adorned themselves with exquisite turquoise ornaments and shiny bronze items, praying for the blessings of benevolent weather and prosperity for the people on their land.

Erlitou was more than just a city; it was, in fact, a metropolitan hub that exerted its authority over a vast surrounding area.

Based on the criteria of civilization mentioned earlier, the Erlitou society had already possessed metallurgy, urbanism, and belief systems. Despite the absence of written evidence at the Erlitou site, its demonstrated cultural sophistication and influence can be considered a threshold into the realm of ancient civilizations. Considering that the subsequent Shang Dynasty continued the dominance of the Erlitou society in the Central Plains, many scholars believe that the earliest China emerged in the Erlitou region.

In comparison to civilizations in the western part of the continent, such as the Babylonian civilization, the earliest China possessed a distinctive essence from the day of its inception. Unlike the ancient civilizations of the Western world, which were heavily influenced by religion, ancient Chinese civilization was centered around ancestor veneration, forming the foundation of its own cultural structure.

How are the differences in civilizations from different regions shaped?

More than 5,000 years ago, in the vicinity of the estuaries of the Tigris and Euphrates Rivers in the Mesopotamian region, a cluster of mud-brick urban cities known as Sumerian cities emerged. These were among the earliest cities on Earth. The Sumerians were the inventors of cuneiform writing, one of the earliest forms of symbolic writing, and the earliest settlers in the Mesopotamian region. The emergence of these Sumerian cities was closely tied to trade, as their locations served as junctions between overland caravans and maritime and riverine trade routes.

Within these Sumerian cities, the population was mainly divided into three categories. The first category comprised local nobility who ruled over the local farmers and slaves, owning the surrounding farmland. The second category included merchants, as well as attendants and sailors who served trade caravans. The third category consisted of priests who also owned land, pastures, and a large labor force. Unlike the nobility, these priests were considered servants of the deities in the mortal realm. Their primary duties included building temples and offering goods to appease the gods. Sumerian priests held significant power, and in the early stages, they even elected leaders responsible for city affairs.

Before the emergence of Sumerian cities, the local region was inhabited by hunter-gatherers. Therefore, the Sumerians initially were "outsiders" who built cities from scratch. They developed irrigation agriculture and engaged in both maritime and overland trade to foster urban growth. As outsiders, the upper echelons of Sumerian society relied on religion to establish authority and unite the population. The ethereal and universal nature of the deities in their religion transcended ethnic and class differences, facilitating the rule of Sumerian elites over both local indigenous populations and new migrants from other areas.

Founding cities through religion and governing through divine authority became the hallmark of Sumerians and their civilization. The foreign priests who emphasized trade and wealth played a crucial role in shaping the civilizations of the Mesopotamian region. This characteristic was inherited and developed by various civilizations across different regions and eras in the Western world.

Examining the emergence and evolution of China's early cultures and civilizations in the context of the Sumerian civilization, we can observe similarities between the Liangzhu Culture in the south and the Sumerian civilization. The Liangzhu Culture seemingly erupted suddenly in the Jiangnan region of Southern China with strong religious undertones. This suggests that the upper echelons of the Liangzhu Culture might also have been outsiders. Upon migrating to the Jiangnan region, they established their own spiritual foundations (possibly bringing advanced jade crafting techniques from distant lands) and managed a significant population and territory.

However, the cultures of the northern Central Plains of China, including the earliest forms of Chinese civilization, took a different path from the Sumerian civilization. Archaeological sites reveal that the process of urbanization in the Central Plains followed a different trajectory. Initially, there were numerous agricultural settlements sustaining tribal communities through nearby farmlands. These settlements gradually grew, leading to the emergence of regional cities with trading functions. Unlike the Sumerian model, China's development began with farmlands and then progressed to cities. The paths of development were distinctly different.

Given that Chinese cities expanded step by step from older agricultural settlements, the phenomenon of "foreign monks reciting scriptures" hardly existed. Belief within tribes was primarily overseen by tribal leaders and their kin. Tribal leaders served as both administrative heads and religious leaders, often taking on the role of ancestral worship or delegating certain roles to their kin.

Within each tribe, the members were closely related by blood, making the worship of common ancestors more important than belief in shared deities. This practice facilitated the rule and unity of tribes by their leaders. Often, the tribe's ancestors were regarded as akin to "deities."

Even as tribes formed alliances and progressed toward the early Chinese civilization, the worship of common ancestors remained a spiritual bond between closely related tribes and within tribes themselves. China's expansive Central Plains were fertile grounds for agriculture, and the emphasis on kinship within the widespread tribal alliances played a key role in shaping early China. Ancestor worship became the spiritual core and belief foundation of early China. Early China was the homeland of its ancestors.

The land nurtures its people. Early China's foundation was a result of the accumulation of local resources and cultural ideas, the influx of foreign goods and exotic cultures, and the collaborative effects of dramatic climatic changes. As China stepped into the realm of ancient civilizations, the fate awaiting our ancestors was not a tranquil pastoral ballad of peaceful years, but a tempestuous collision of civilizations.

The Column: Buried within the Tooth Decay, Lies the Taosi People's Culinary Secrets

Modernity prompts us to tend to our teeth with "religious" regularity, a practice endorsed by dental sages who advocate brushing after every repast. The foremost aim: ward off the specter of cavities. Known colloquially as "虫牙" (chóng yá) in Chinese, these oral adversaries are nurtured by the sweet whispers of sugar found within our victuals.

In antiquity, dental hygiene suffered a graver fate. Carbohydrates, the core components of yesteryear's fare, metamorphosed into acids courtesy of insidious bacteria post-consumption. These corrosive agents, in turn, eroded teeth, giving birth to the infamous cavities. Scholars, both foreign and native, have laboriously computed cavities' historical incidence, shedding light on civilizations past.

Distinguished researchers reveal that among our ancestors, who traversed the terrains as hunters and gatherers, cavities prevailed at a mere 0 to 5.3%, averaging at 1.3%. Those embracing a mélange of hunting, gathering, and agriculture encountered a moderate cavities presence ranging from 0.44% to 10.3%, with a mean of 4.8%. The farmers of old, however, bore the brunt, witnessing the cavities' surge at 2.1% to 26.9%, averaging 8.6%. Yet, there's a revelation: a nomadic lifestyle safeguarded against this dental affliction, beckoning only 0.2% to 0.9% cavities.

Thus, it becomes apparent—carb-laden agriculture primes cavities for conquest.

Such revelations serve as a window into the societies of yore. Enter the saga of the Taosi Culture, an antiquarian narrative flourishing 4,300 to 3,900 years in the past—concurrent with Shaanxi's Shimao Culture. The excavations at Taosi unearth tombs and ash pits, whispering tales of humanity. By assessing the relics they lay with and the architecture that embraced them, one infers the tomb residents' elevated station while the ash pit's denizens resonate with commonality. Botanical relics divulge an advanced agrarian landscape where foxtail millet held dominion. The skeletal remains of animals extracted from the earth weave a story dominated by mammals—pigs, cattle, sheep, dogs—the swine ranking foremost. This petrified menagerie suggests that meat sourced from tamed creatures formed the crux of their carnivorous pursuits.

And what was the epicurean chronicle of the Taosi folk? By gauging the oral vestiges left behind, scholars discerned a cavities rate near 3%— tombs at 2.9% and ash pits at 3.1%, barely a smidge's difference. Evidently, the Taosi populace, despite mastering the agricultural symphony, serenaded their palates with a diverse melody—meat, notably, joining the culinary chorus. Experts delved further, scrutinizing dental abrasion, unveiling a moderate attrition rate, a phenomenon entwined with their consumption of tender meats. The scrutiny of microelements within bones hinted at a diet unswervingly tethered to the consumption of meat, a gustatory anthem that resonated across epochs.

Taosi Culture's narrative exhorts the notion that agricultural mastery need not denote monotonous staples. While the verdant harvest furnished energy, its gustatory and nutritional opus was lackluster. The ancients, thus, enriching their repasts with carnivorous interludes, fashioned tables of indulgence when alternative nourishment—particularly meats—beckoned.

Another tableau beckons, the epoch of the Hongshan Culture an ancient saga predating Taosi's. Hongshan's cavities waltz measured between 1% and 2%, a cadence resonating with perhaps modest agrarian exploits. Yet, it echoes a refrain of diversified sustenance, courtesy of the neighboring wilderness and the thrall of the hunt.

The twilight of Taosi's splendor unfolded abruptly—turmoil and upheaval danced through its final days. Palaces and ramparts crumbled, noble resting places desecrated, lives sacrificed. Scholars duel, debating whether this dirge was composed by internal dissent or external invasion. Amidst this speculative tempest, cavities offer a lens into a milieu of culinary concord—a testament that hierarchies did not sow discord through their culinary distinctions.

Hence, these diminutive cavities emerge as silent sentinels, sharing whispers of antiquity's gastronomic symphony—a trove of wisdom from ages past.

CHAPTER 5

Ascendancy in the Western Realm: Unveiling the Cascading Dominos Toward the Eastern Vistas

> In regal raiment, a dark mane cascading down the center, crowned by an apex of refinement embellished with a crimson ribbon, her gaze softly veiled, as if stirred from a reverie, a graceful aquiline nose, lips slightly agape, a glimpse of teeth, all bestowing an everlasting smile upon posterity.

Archaeologist Hans Folke Bergman (1902–1946) thus narrates his encounter with the time-honored sepulcher's enigma, the feminine relic dubbed the "Smiling Princess" (微笑公主) of Xiaohe Site. In the tapestry of 1934, Bergman, in the bloom of his thirty-first year, was introduced to the "Smiling Princess," a sentinel of 3,800 years elapsed—a temporal memoir of her corporeal existence.

Slumbering amid the desert's embrace, the "Smiling Princess," entwined with the ancestral essence of her companions interred within the Small River Cemetery, beckons the gusts of history to disrobe the ages, unveiling the odyssey of human migration across the Western dominion, spanning millennia.

5.1 The Complex Lineage of the Princess of Xiaohe (小河公主)

Folke Bergman was Swedish. When he graduated in archaeology in 1927 and came to China, joining the China Northwest Scientific Expedition led by Sven Hedin (1865–1952), his Swedish predecessor Johan Gunnar Andersson had not only already discovered the Yangshao Culture site but had also just announced to the world the discovery of teeth from the Peking Man at Zhoukoudian in the previous year, establishing his reputation.

With the help of locals, the Northwest Scientific Expedition discovered a group of ancient tombs near Lop Nur in the Tarim Basin of Xinjiang, China, containing "a thousand coffins." Bergman casually named this burial site "Xiaohe Cemetery" after a nearby small river, which is where the Princess of Xiaohe rests.

A glimpse spanning thousands of years turned into a farewell. Due to the tumultuous social unrest in China at that time, Bergman left Xiaohe Cemetery with a small collection of artifacts and never returned. The Princess of Xiaohe appeared astonishingly but then vanished into the endless sands.

After over 60 years of fervent searching, in the year 2000, the Xiaohe Cemetery finally reappeared before the world. This time, archaeologists discovered the mummified remains of a young woman. She wore a pointed felt hat and had slightly closed eyelids, her long eyelashes guarding her beauty across the ages. She came to be known as the Princess of Xiaohe, corresponding to Bergman's description of the "Smiling Princess" (微笑公主).

According to radiocarbon dating, the Xiaohe Cemetery dates back to approximately 3,540 to 3,980 years ago. This likely indicates that the Princess of Xiaohe and her people were among the earliest settlers in the Lop Nur region. Analyzing the genetics of this early population could unveil a chapter of prehistoric history in the Tarim Basin.

Genetic analysis reveals that the maternal lineage of the Xiaohe Cemetery population carries genes both from the Western Eurasian continent and the Eastern Eurasian continent. However, this doesn't imply that Eastern and Western populations had mixed in the Lop Nur region nearly 4,000 years ago. A more detailed genetic analysis suggests a different narrative: Eastern Eurasian populations, possibly from the Mongolian Plateau and Baikal Lake regions, likely met Western populations on the Eurasian steppe, where they merged, possibly in the southern part of Siberia or the grasslands of Central Asia, such as Kazakhstan. Subsequently, this mixed population migrated southward, with one group settling in the Lop Nur region.

This is the origin story of the Princess of Xiaohe population. Hence, they could be descendants of pastoralist populations from the northwest or north.

Furthermore, the Xiaohe Cemetery population also carries a small percentage of genes from South Asia and the Central Asian Iranian region. One explanation is that a group of South Asian populations traversed the treacherous Pamir Plateau, entered the Tarim Basin, and then continued eastward to the Lop Nur region. This route is quite challenging, so another explanation might be more plausible: Some South Asian genes initially merged with the populations of the Eurasian steppe, then migrated southeastward into the Northern Xinjiang grasslands, and finally crossed the Tianshan Mountains into the Lop Nur region.

The origin of the population in the Xiaohe Cemetery is complex, and over time, it may have incorporated individuals from various sources. In the early stages, the genetic diversity of the Xiaohe Cemetery population was low. For instance, a particular maternal lineage genotype accounted for approximately 50% of the entire population. This included the genetic profile of a female corpse found within a ritually buried mud coffin. Her genetic makeup belonged to this lineage, suggesting her high status and likelihood of having borne many offspring. It's possible that the initial migrating group was essentially a large family, with close blood ties among female members. However, the prevalence of this genetic type decreased rapidly in later periods, and the phenomenon of mud coffins disappeared, indicating an increasingly diverse origin of women living in the region of Lop Nur.

On the other hand, the paternal genetic lineage of the Xiaohe Cemetery population presents a different picture, predominantly consisting of genetic types from the Western Eurasian continent. Moreover, there isn't a significant difference between early and late periods. This suggests that they may not have accepted or had very few external male members, instead continually incorporating females from outside. It's plausible that these female members were brought in through marital unions, and it's even possible to speculate that many Lop Nur women married into surrounding communities.

The Xiaohe Cemetery population retained many cultural elements from the grasslands. For instance, their burials included many cattle skulls and sheep horns. Graves contained the buried remains of bulls and rams, with some tombs containing as many as 26 of these items. Coffins were covered with whole cowhide or sheepskin, indicating that these individuals were likely involved in animal husbandry. While no pottery was found in the Xiaohe Cemetery, similar sites nearby yielded pointed-bottom baskets that closely resemble a style of pottery common on the grasslands.

The Xiaohe Cemetery population included wheat among their grave goods, making it one of the earliest sites in China where wheat has been discovered. Additionally, remains of foxtail millet and proso millet grains were unearthed. This combination of crops in a complementary style likely emerged along the grassland route. Therefore, it's conceivable that the ancestors of the Xiaohe Cemetery population engaged in East–West agricultural exchange along this route, bringing diverse crops to the Lop Nur region during their southward migration.

An even more intriguing discovery is milk. Residues of a milky-white substance were found in burial baskets, on the clothing, and on the skin of the deceased individuals. Protein analysis confirmed that this substance was derived from milk products, indicating that milk had become a significant part of the dietary habits of the Xiaohe Cemetery population. Western Eurasian populations possess lactase persistence genes, allowing them to digest milk, while Eastern Eurasian populations lack this gene. The phenomenon of consuming milk observed among the Xiaohe Cemetery population, including the Xiaohe Princess and her group, once again links them to Western Eurasian populations. The Western Eurasian genes that the Xiaohe Cemetery population possessed might have enabled them to digest lactose, allowing them to consume milk without issues.

Whether through genetic analysis or cultural comparisons, the Xiaohe Cemetery population's origins point to the Eurasian grassland belt, specifically an ancient grassland culture known as the "Qiemu'erqieke Culture."

The earliest agriculture in human history originated in the Western Asia, and animal husbandry emerged after surplus grains were available for feeding and domesticating animals in agriculture. Some farmers brought domesticated animals to the vast grasslands and became pastoralists, specializing in animal husbandry, while others engaged in agricultural farming, thus forming a division of labor. Not surprisingly, the earliest grassland culture in the Eurasian steppe belt originated from the western part of the Eurasian continent, as it was closer to the birthplace of agriculture and the division of labor between farmers and pastoralists occurred earlier.

Around 5,600 years ago, in the western part of the grassland belt, likely on the steppes in what is now southern Russia, the Yamnaya Culture emerged. These people of the grasslands mastered bronze technology, domesticated horses, and likely invented the wheel. A prominent cultural feature of theirs was the kurgan burial (竖穴墓), which involved digging burial chambers directly beneath the ground's surface. Genetically, both the Yamnaya population and various later nomadic groups on the steppe carried genetic components related to the surrounding agricultural populations, indicating their differentiation from agricultural groups.

From genetic research, over 4,000 years ago, descendants of the Yamnaya Culture had spread to the eastern part of the Eurasian steppe belt. For example, ancient human remains found in the Baikal Lake region contained genetic components from the Western Eurasian steppe, and the populations that succeeded the Yamnaya Culture in the southern Russian steppe already carried genes from populations in the Eastern Eurasian steppe. This suggests that genetic exchange between Eastern and Western steppe populations started quite early.

Even more intriguingly, in ancient human remains from the Baikal Lake region dating back 4,500 years, molecular biologists detected genetic information of an ancient strain of Yersinia pestis, the bacterium responsible for the plague. This Yersinia pestis strain likely originated among the Yamnaya people in the Western Regions and then spread to various areas as these populations migrated. However, the two individuals from the Baikal Lake region with detected Yersinia pestis do not carry the unique genes of the Yamnaya Culture. This leads to the speculation that Yersinia pestis might have "preceded" the genetic lineage, spreading to the Eastern Eurasian steppe even before the Yamnaya genes did, through infections among different populations.

During this era of the expansive and dynamic steppe cultures, the Qiemu'erqieke Culture emerged near the Altai Mountains in Northern Xinjiang, which is in Western China.

From the perspective of population genetic exchange, the Qiemu'erqieke people likely inherited not only the Western genes from the Yamnaya Culture but also incorporated genes from the east, such as those of ancient Siberian populations. This could explain why individuals from the Xiaohe Cemetery population also exhibit complex genetic profiles, as their Qiemu'erqieke ancestors had already mixed the bloodlines of both Eastern and Western populations.

The Qiemu'erqieke people primarily thrived in the grasslands between the Altai and Tianshan Mountains. The Qiemu'erqieke Culture represents one of the earliest Bronze Age Cultures in Xinjiang. Archaeologists have discovered that the Qiemu'erqieke people were skilled in carving and smelting bronze using stone crucibles. An excavation at a Qiemu'erqieke cultural site in Qitai County, Xinjiang, unearthed a stone crucible with a diameter of nearly 20 centimeters and a height of nearly 40 centimeters, complete with a round handle.

Remarkably, the style of this stone crucible is strikingly similar to a pottery crucible from the Shang Dynasty, known as the "General's Helmet," unearthed in Anyang, Henan province. One of the differences lies in the material—the Anyang crucible was made of clay pottery, while the Qitai County crucible was carved from stone. Another distinction is the timeline—the Anyang pottery crucible dates back several centuries later than the stone crucible located over 2,000 kilometers away.

Due to the advanced metallurgy of the Shang Dynasty, which did not pass through the natural copper stage but directly adopted bronze metallurgy, it is likely that their bronze metallurgical skills originated from West Asia and spread into the Central Plains via the Eurasian steppe. This Qiemu'erqieke stone crucible provides evidence for the west-to-east transmission of bronze technology.

5.2 The Twin "Wings" of Destruction: The Dance of Grassland "Butterflies"

Not long after the Qiemu'erqieke people, who blended Eastern and Western influences, rose to dominance on the Northern Xinjiang grasslands, a drastic change swept through the Eurasian steppe. To the west of the Qiemu'erqieke Culture, a formidable power emerged, characterized by the use of horse-drawn chariots for warfare. This force advanced eastward, posing a significant threat to the Qiemu'erqieke people.

Furthermore, in the northern reaches of the Qiemu'erqieke Culture, another powerful group armed with single-hook spears (单钩长矛) and bronze socketed axes (空首铜斧) expanded southward. Faced with pressure from multiple strong adversaries, the Qiemu'erqieke people had to yield and migrate southward and eastward. One branch crossed the Eastern Tianshan Mountains and settled in the Lop Nur region, becoming the population of the Xiaohe Cemetery.

The Asian northern force wielding spears and copper axes that pushed the Qiemu'erqieke people away was the Okunev group. They originally thrived in the Minusinsk Basin, situated in the middle reaches of the Yenisey River in Siberia. The name "Okunev" comes from the village where this culture was first discovered. It's worth elaborating on the Minusinsk Basin—a region located in the Southwestern Siberian area near the northwestern borders of China's Xinjiang and Mongolia. The basin features diverse landscapes such as mountains, hills, grasslands, and valleys, with abundant natural resources like forests, water, and mineral deposits that supported human survival.

The relatively secluded natural environment made the region defensible, resulting in fewer external disturbances. During the Bronze Age, it served as the cradle for significant steppe cultures. Notably, the Minusinsk Basin even had cultural interactions with distant regions like the Central Plains.

As a part of the Eurasian steppe complex, the Okunev group's genetic makeup also blended genes from both Eastern and Western populations. Their primary economy revolved around animal husbandry, supplemented by fishing and hunting. The Okunev group's ability to subdue the Qiemu'erqieke people might have been due to their more advanced bronze technology. They employed sophisticated tin-bronze casting techniques, crafting their red copper axes through casting as well. This resulted in highly sharp single-hook spears and bronze socketed axes.

The exceptional weaponry of the Okunev group was coveted treasures across the Eurasian steppe, hence similar styles of weaponry were discovered in many regions across Eastern Europe and Asia. These weapons were found not only in the Altai Mountains but also to the west in the Urals at the crossroads of Europe and Asia, where four Okunev spears were unearthed. Further east, in Gansu province's Machang Culture (马厂文化) site, two spears were also found. In slightly later cultures in Gansu, such as the Qijia and Siba Cultures, hollow-handled copper axes were discovered.

However, the widespread distribution of Okunev weapons didn't necessarily signify that the Okunev group conquered lands and established a vast steppe empire. They hadn't reached that level of power. In reality, both the Qiemu'erqieke people and the Okunev group encountered staunch resistance as they attempted to expand eastward into the heartland of China. The formidable opponent they faced was the Siba Culture.

The Siba Culture thrived around 3,900–3,500 years ago in the western part of Gansu province to the eastern part of Xinjiang. To the east, it bordered the Qijia Culture, and further east, the Central Plains Cultures headed by the Erlitou Culture.

The people of the Siba Culture are believed to be the main ancestors of the Qiang people, an important ethnic group in Northwest China. Genetically, they are primarily of East Asian origin. Due to their proximity to the steppe and Central Asia, these ancient Qiang tribes were exposed to the Bronze Age Culture early on. They extensively cultivated and domesticated various plants and animals imported from the steppe region. In the arid environment of Northwest China, these tribes grew barley and wheat, while also herding sheep, goats, cattle, and horses. The Siba Culture adopted many weapon technologies from the west, possessing bronze weapons such as adzes, spearheads with side blades, and three-edged guard swords. Many of their weapons prototypes can be found in the steppe region.

Interestingly, their culture also included artifacts like staff heads, a characteristic cultural element from the Western world symbolizing power. Ancient Egyptian civilization widely used staff heads, and they were also popular in regions of West Asia. The earliest staff head discovered in China was found in the Tianshui region of Gansu province. The distribution of staff heads in the northwest of China displays an east-to-west pattern, with staff heads appearing later in Xinjiang than in Gansu. This artifact's origin from the West further emphasizes the transmission of Western products, technologies, and culture along the steppe belt into the heartland of China, often following a route through the Mongolian Plateau.

It was precisely the ancient Qiang people who inhabited the Western Gansu and Eastern Xinjiang regions, extending their influence westward into the Hami Basin. They steadfastly resisted the eastward expansion of Altaic tribes represented by the Qiemu'erqieke Culture and Okunev Culture, which prevented the powerful Western Eurasian force, characterized by a fusion of East and West genetic types, from advancing beyond Xinjiang. Otherwise, many cultural groups on the Chinese land might have integrated more extensively with the genetic makeup of the Western populations of the Eurasian steppe.

However, during the Bronze Age, the steppe regions were in turmoil. Bronze weapons and horse-drawn chariots had become the standard equipment of dominant groups. Wave after wave of steppe tribes mounted on horseback, fiercely contested for territory and wealth, disrupting the tranquility of the Chinese land.

Around 3,400 years ago, within the Minusinsk Basin, another formidable steppe ethnic group emerged—the Karasuk people. After their rise, the Karasuk expanded southward to relatively warmer regions, entering the northern grasslands of Xinjiang in China and becoming a significant influence in the northwest. Genetically, the Karasuk population carried both the genetic traits of western steppe populations and those of indigenous populations from Southwestern Siberia.

What's particularly intriguing about the Karasuk Culture is its technological and cultural exchange with the contemporaneous powerful Shang Dynasty in the Central Plains. The elbow-shaped bronze knives (曲柄青铜刀), a standard tool of the Karasuk Culture, is quite similar to certain styles popular in Northern

China. The pottery style used by the Karasuk people also bears resemblance to the popular styles in Northern China.

An analysis of the evolution of bronze artifacts reveals that there was indeed interaction between the Karasuk Culture based in the Minusinsk Basin and the Central Plains region. Karasuk weaponry such as grooved swords (凹格剑), animal-head swords (兽首剑), bell pommel swords (铃首刀剑), bow-shaped ornaments (弓形器), etc., were influenced by Northern China. A small number of three-legged egg-shaped vessels (三足蛋形瓮) were found in the Minusinsk Basin, which first appeared in Central Shanxi, China. Conversely, many animal-head style swords in Northern China might have been influenced by the Karasuk Culture.

Why did such distant grassland powers and Central Plains powers show such apparent cultural exchange?

The Karasuk Culture spanned the steppe for 500 to 600 years, covering most of the Shang Dynasty and the early Zhou Dynasty Periods. During the Bronze Age, bronze was the cutting-edge technology and a strategic resource of the time. Relatively speaking, the Shang Dynasty, situated in the Central Plains, possessed unparalleled bronze ore resources and production, along with unmatched metallurgical techniques in East Asia. A powerful steppe power naturally desired to establish connections with the Shang Dynasty, a "technological and resource-rich nation," in order to acquire bronze artifacts and metallurgical knowledge, even if the two were separated by thousands of kilometers. Moreover, it's likely that the Shang ethnic group migrated from northern regions to the Central Plains, maintaining deep connections with the northern areas and even the steppe regions.

Ironically, the flourishing of the Karasuk Culture indirectly contributed to the downfall of the Shang Dynasty.

The ability of the Karasuk people to interact with the Shang Dynasty implies that the space between the two regimes was already populated by diverse ethnic groups, serving as intermediaries for long-distance cultural exchanges. Whether in Northern China or the Mongolian Plateau, numerous ethnic groups coexisted, engaging in either conflict or cooperation. The expansion of the Karasuk people inevitably triggered a cycle of large-scale inter-ethnic conflicts and migrations.

Just like a row of dominoes, knocking down the first domino will trigger a series of others falling. As the Karasuk people drove their chariots south and east, the Altaic tribes that originally lived on the Northern Xinjiang grasslands sought survival to the southeast under pressure. This pushed against the ancient Qiang tribes inhabiting the Eastern Xinjiang and Western Gansu regions. The Qiang tribes then transmitted this pressure further east along the Hexi Corridor, influencing various ethnic groups in Eastern Gansu and Shaanxi. Among these ethnic groups were the Zhou people.

Historical records indicate that the Zhou people initially lived "between the Rong and Di people" (currently in the Qingyang City in Gansu province). According to the Shang Dynasty's classification of external ethnic groups, those to the west were called Rong (戎), and those to the north were called Di (狄). The Zhou people happened to be situated in the northwest, possibly active in the area from Eastern Gansu to Western Shaanxi.

The historical records further state that the early leader of the Zhou people, Ancient Duke Danfu (古公亶父), led his tribe to migrate from the region of Bin (豳) to the open plains below Qishan due to threats from the Rong and Di tribes. This event is known as the "Danfu's Migration to Qishan," marking the start of a new chapter in Zhou history as they gradually grew in the region below Qishan.

Subsequently, the Zhou people advanced to the east. During a time when the Shang Dynasty was exhausted from fighting the Dongyi people and vulnerable in the rear, the Zhou people, armed with bronze weapons obtained through interactions with the Rong and Di, as well as chariots pulled by horses, launched a surprise attack on the Shang Dynasty's capital city of Shanggao. In a single battle, they seized power, overthrew the mighty Shang Dynasty, and established their own dynasty by conquering a "greater state" (大邦) with their "lesser state" (小邦).

The historical records reveal the secret of the massive ethnic migration during that time. From the Karasuk people, Altaic tribes, Qiang tribes, and Zhou people to the merchants, military pressures were released from the Minusinsk Basin in Southwestern Siberia, transmitted layer by layer, and ultimately pressed toward the Central Plains. The Karasuk people from Southwestern Siberia brandished their bronze swords across the mountains, and the 600-year Shang Dynasty in the Central Plains collapsed.

A butterfly lightly danced on the grasslands, and thousands of miles away, a storm of bloodshed and change of dynasty swept across the Central Plains.

After toppling the Shang Dynasty, the Zhou people inherited the Bronze Culture and inscribed their ancestors' greatness and their own magnificence on bronze inscriptions. However, regardless of how the Zhou people embellished their early history on bronze artifacts and claimed moral superiority, it cannot be denied that their ethnic origins were also rooted among the Rong and Di tribes. The fact that they sought survival space to the east under military pressure from the west cannot be denied either. Fortunately, their eastern expansion succeeded.

From the Qiemu'erqieke people migrating south to Lop Nur to the Zhou people overthrowing the Shang Dynasty, over the span of approximately 1,000 years, the fundamental trend of "Western Regions Hegemony" was that grassland tribes held the advantage. These tribes often hailed from the western or northern grasslands of the Siberian region, continually expanding their influence to the east and south. Faced with the invasion of formidable enemies, evading was not the solution; the Zhou Dynasty needed strategies to both fend off external threats and maintain internal stability.

5.3 Residing in China, Defending the Neighbors

When tracing the earliest appearance of the term "China" (中国, zhōng guó), one will mention a national treasure-level bronze artifact—the He Zun (何尊), along with the inscription on it—Here in China I Dwell (宅兹中国).

After achieving a decisive victory in the overthrow of the Shang Dynasty, King Wu of Zhou pondered how to govern such an extensive territory and numerous population. He confided to his minister Dan, Duke Wen of Zhou (周公旦), "We haven't yet established a capital city, and I can't sleep through the night." Observing the heavenly patterns, he believed that the region of Luo (洛) was not far from the heavenly capital (上天都邑), and a city should be built there in the future. This story is recorded in the pre-Qin ancient text "Lost Book of Zhou" (《逸周书》).

Only two years after defeating the Shang Dynasty, King Cheng of Zhou, Ji Song (姬诵), ascended the throne after the death of his father King Wu. His mother was the daughter of Lv Shang (吕尚, or Jiang Shang), also known as Jiang Ziya (姜子牙), the protagonist of the Chinese legend that "fishes rising to Grand Duke Jiang Ziya's hookless and baitless line" (姜太公钓鱼——愿者上钩).[1] King Cheng followed his father's wish and began constructing the capital city in Chengzhou (洛地) while gathering clans for instructions. A nobleman named He cast a bronze vessel, and on the inscription at the bottom of the vessel, a commemoration of this significant event was carved. This is the origin of the He Zun.

---

1 "Jiang Taigong Fishes" is a legend from ancient China about Jiang Shang/Lv Shang/Jiang Ziya, an old fisherman who despises the tyrannical rule of the last ruler of the Shang Dynasty. He patiently waits for an opportunity to help overthrow the king. King Wen of the Zhou state discovers Jiang Shang and realizes he is a wise political thinker and military strategist. King Wen appoints him as prime minister and renames him Jiang Taigong. Jiang Taigong advocates for a ruler to prioritize the prosperity of the people and reduce taxes and labor. After King Wen's death, his son King Wu seeks to overthrow the Shang king, but Jiang Taigong advises patience. When the people of Shang lose faith in their ruler, Jiang Taigong and King Wu launch an attack. A decisive battle is fought at Muye, resulting in the downfall of the Shang Dynasty and the establishment of the Zhou Dynasty. Jiang Taigong plays a crucial role in the victory and is later made the duke of the State of Qi, which thrives under his leadership.

Within the 12 lines and 122 characters of the He Zun's inscription, the phrase "Here in China I reside, from here I pacify the people" (余其宅兹中国，自兹乂民) appears, marking the first appearance of the term "China" in the river of history.

Scholars are still debating the interpretation of the characters "中国" (China) in the He Zun inscription. However, the general consensus is that this bronze vessel, crafted in the early Western Zhou Period, does not convey the modern understanding of "central country." Some scholars believe that the term "中国" in the inscription refers to Luo's convenient location, situated advantageously in the center of the realm. Others posit that the character "中" in "中国" represents the ancient drum used in rituals and wars, a type of drum elevated by a tall stand.

Even today, ethnic minorities like the Miao people (苗族) in Southwestern China continue to use similar drums in funeral ceremonies. Therefore, the meaning of "宅兹中国" can be understood as "residing in Luo, conducting rituals in Luo."

Moreover, analyzing the sentence structure, "宅兹中国" and "自兹乂民" should be of the same structure. The character "乂" signifies governance and stability, thus the preceding "中" should also have a verb meaning, representing rituals. During the Shang and Zhou Periods, maintaining spiritual unity through rituals was of utmost importance. Considering King Wu's choice of Luo based on Zhou's celestial beliefs for the capital's location, interpreting "中国" as "conducting rituals in Luo" might be more plausible.

Therefore, the He Zun inscription reflects the internal construction of the upper echelons of the Zhou Dynasty after overthrowing the Shang Dynasty. At that time, the dynasty had just undergone a drastic change in leadership, and the domestic situation remained unstable. The line in the inscription, "Only King Wu has conquered the great city of Shang" (唯武王既克大邑商), reflects the hard-won victory, akin to swallowing an elephant—an indication that consolidating the victory was no easy task. The Zhou royal family had to find ways to bind the nobles tightly together.

The Zhou Dynasty greatly promoted the foundation of Chinese civilization—ancestor worship tradition. They established the Patriarchal clan system (宗法制度), with distinctions between the emperor (天子), feudal lords (诸侯), ministers (卿大夫), and scholars (士). Within clans, there were divisions based on seniority, with major and minor branches, allowing nobles to understand their status. Depending on their rank, they enjoyed different levels of rituals.

This how the interior security was made in the Zhou Dynasty. While securing the interior, the Zhou Dynasty also needed to "repel external threats," defending against increasingly intense invasions from the surrounding regions, and clearly delineating the boundaries between their own ethnic group and country, and those of foreign ethnic groups and countries.

In the later stages of the Shang Dynasty, merchants already sensed the threat posed by northern and western ethnic groups. The borders of the Shang Dynasty were frequently harassed by enemies. For instance, one oracle bone inscription records a day when the "Gong Fang" (工方) invaded the Shang Dynasty, plundering the population. Another day, a diviner asked the Shang king if the future would be peaceful. After examining oracle bones, the king determined that the situation was not favorable.

A few days later, news arrived that the "Tu Fang" (土方) had invaded the Shang Dynasty from the west, occupying two towns. Another major opponent of the Shang Dynasty was the "Qiang Fang" (羌方) to the northwest. In the earliest oracle bone inscriptions, it's recorded that the Shang Dynasty sacrificed Qiang prisoners. Undoubtedly, the Qiang people did not spare the Shang Dynasty either, and the conflicts between the two sides persisted. The once weak Zhou people had once sought refuge with the Shang Dynasty and assisted them in dealing with these formidable enemies from the west.

Through oracle bone inscriptions, the Shang people documented the various ethnic groups that posed significant threats to their country. However, beyond these inscriptions, the Shang people might not have fully understood the profound social changes occurring in the Bronze Age grasslands. These changes prevented both the Shang and subsequent Zhou people from sleeping peacefully.

During that time on the grasslands, with the domestication of large animals like horses, cattle, and camels completed, the grassland communities gained a deeper understanding of the grassland's characteristics. This marked the transition from pastoral economy to nomadic economy.

Typically, a pastoral economy was often just a "side job" or one of several "main jobs" for agricultural communities, which still relied on an agricultural economy. For example, archaeologists discovered that 6,000 years ago in the northern mountains of Iran, people primarily engaged in irrigation agriculture while also raising goats and sheep, supplementing their farming economy. Moving west to the Caucasus region, farming and animal husbandry held equal importance, with cattle, sheep, and pigs as the main livestock. Moving further west into Eastern Europe, people there engaged not only in farming and animal husbandry but also retained hunting and gathering for subsistence.

Although nomadic economy evolved from pastoral economy, it had a fundamental difference—it could function independently from agricultural economy. Nomadic communities roamed the grasslands with their livestock as the main or even sole occupation. Pastoralists had limited activity areas, often revolving around their small agricultural settlements. In contrast, nomads had vast territories, rotating grazing areas with the seasons. This ensured that livestock had abundant, fresh grass to eat and that the grasslands thrived due to the animals' trampling and manure.

This is the intriguing and unique aspect of grassland ecosystems. Biologists understand that forest ecosystems thrive with minimal human intervention, while agricultural ecosystems heavily depend on human intervention. Unlike the former two, the prosperity of grassland ecosystems requires human intervention but not excessive interference. It's a system where grass, livestock, and nomads coexist—they mutually influence one another, achieving dynamic balance through nomadic economic activities.

Nomadic communities gradually understood the laws of the grasslands. For instance, more than 3,000 years ago in the Western Eurasian grasslands, pastoral communities would migrate and expand every 20 years or so because the local grazing lands deteriorated after years of use.

These communities were likely the earliest practitioners of nomadism. The nomadic economy might have initially emerged within a small scope of the grasslands, but as nomadic communities emerged, their demand for grazing space drove expansion. To sustain themselves during constant migration, they would intrude upon neighboring settled pastoral and agricultural societies, posing a serious threat to both pastoral and farming communities.

This process forced many communities to arm themselves, leading to differentiation: some increased their mobility, becoming part of the nomadic groups, while others enhanced settlement defenses, relying more on the advantages of agricultural society—food, population, and technology—to confront the nomads.

In this way, the two major camps of the ancient world—agrarian societies and nomadic societies—gradually emerged amid this turbulent situation. Looking at a global scale, around 4,000 to 3,000 years ago marked the "first major impact of the nomadic world on the agrarian world." In the past, there was the theory of the Aryan expansion. The Aryans were a nomadic group living in the Eurasian steppe, who invaded ancient India from Central Asia around the fourteenth century BCE, giving rise to the Vedic culture in Indian history. They also invaded ancient Greece around 1200 BCE, leading to the downfall of the Mycenaean civilization.

In reality, the expansion of these nomadic groups might not have been the work of a single ethnicity. The diffusion of nomadic economies on the steppe led to conflicts and migrations among various groups. From the perspective of the agrarian world, it seemed as though countless horse-mounted ethnic groups emerged on the steppe, threatening to eradicate their cultures.

The expansion of nomadic groups was likely connected to the "chariot revolution" on the steppe. Around 3,700 years ago, lightweight and sturdy horse-drawn chariots entered the historical stage in the northern regions of the

Tigris and Euphrates River Valleys. These chariots could carry a driver and an archer onto the battlefield simultaneously, combining mobility and firepower and fundamentally altering the warfare of the ancient world. Scholars once collectively referred to the groups driving these horse-drawn chariots out of the steppe as Aryans.

Soon, the horse-drawn chariots spread across the Tigris and Euphrates River Valleys, then westward into ancient Babylon and eastward into ancient India. Hundreds of years later, the influence of horse-drawn chariots expanded into Scandinavia in Northern Europe and the northern regions of China, during the later stages of the Shang Dynasty.

In the first half of the twentieth century, during the rise of racism, Nazi Germany propagated the idea that the noble Aryan race was the origin of the Germanic people, drawing on theories from scholars like Ernst Haeckel, mentioned in the first chapter of this book. However, genetic analysis shows that modern Germanic people only inherit about 20% of Aryan genes. The modern populations closest genetically to the ancient Aryans are Russian, Ukrainian, and Polish people—a fact ironically contradictory to the Nazis' racist beliefs.

This genetic distribution can be understood easily: the Aryans were a nomadic steppe group, so their closest relatives genetically would be those with more steppe ancestry rather than agricultural ancestry.

Now, let's return to the East and the Shang and Zhou Dynasties. The Zhou Dynasty's rise in place of the Shang Dynasty was essentially a result of the expansion of nomadic groups across the Eurasian continent. The Shang Dynasty was primarily agrarian, while the Zhou Dynasty, after replacing the Shang, transformed from a northwestern power with mixed agriculture and herding to a predominantly agrarian dynasty in the Central Plains. The wave of disruption caused by nomadic groups was a challenge that both dynasties had to face.

The fall of the Western Zhou Dynasty was also directly related to the threat of nomadic groups. We are familiar with the story of "The Boy Who Cried Wolf"—in this case, King You of Zhou (周幽王) lit the alarm fires to entertain his favorite concubine, Bao Si (褒姒). He repeatedly mocked the lords who came to rescue, and later, the Quanrong (犬戎) nomads actually attacked the Zhou capital. Ignoring the alarms, the lords failed to provide aid, resulting in the fall of the capital and the demise of the Western Zhou Dynasty.

This Eastern version of the story carries many elements of fiction, but some information might align with historical facts. To defend against the attacks of nomadic groups that moved swiftly like the wind, the Zhou Dynasty would have established an early warning system. Nearby lords would form a joint defense system, ready to rush to aid in case of emergency. Historians currently believe that the Quanrong were brought in by one side during an internal conflict within the Zhou Dynasty.

The people known as Quanrong (犬戎) in historical texts, also referred to as "Xianyun" (猃狁), were a nomadic group active in the regions of Gansu and Shaanxi. The dominant presence of these groups in the northwest region lends support to the idea that the Zhou people's eastward expansion and the overthrow of the Shang Dynasty were indeed a process of "pressure propagation." It wasn't the Zhou people's strength that led them to occupy the northwest; rather, they were pushed eastward by other ethnic groups in the northwest. Even after the establishment of the Zhou Dynasty, the Quanrong remained a nightmare for the Zhou people.

In the "Book of Songs—Cai Wei" (《诗经·采薇》[2]), there is a passage:

> Let us gather the thorn-ferns, let us gather the thorn-ferns;
> The thorn-ferns are now springing up.
> When shall we return? When shall we return?
> It will be late in the [next] year.
> Wife and husband will be separated,
> Because of the Xian-yun.
> We shall have no leisure to rest,
> Because of the Xian-yun.

This poem depicts the grievances of the border guards of the Zhou Dynasty. Peasants gathered peas year after year, saying they would return home, but they never could because of the formidable enemy, the Quanrong.

Looking at a map of China's topography, we see a semi-circular mountainous region surrounding large plains in the east and south. This mountainous region is formed by the Yan Mountains, Yin Mountains, Helan Mountains, and the Hengduan Mountains. During the era of pastoral economy, the cultural differences among various groups in the northern regions of China were not significant. In this mountainous region, people engaged in both agriculture and animal husbandry, as pastoral economies heavily relied on agricultural activities. Therefore, the mountainous communities had deep cultural connections with the central plains agricultural communities. Even the Shimao Culture, situated at the intersection of the Mongolian Plateau and the Loess Plateau, exhibited many characteristics of an agricultural society, emphasizing farming as the primary

---

2 采薇采薇、薇亦作止。
曰归曰归、岁亦莫止。
靡室靡家、猃狁之故。
不遑启居、猃狁之故。

occupation while also practicing animal husbandry and constructing fortified settlements.

However, with the rise of the pastoral economy in the Eurasian steppe, highly mobile grassland groups emerged. They moved swiftly, possessed formidable long-range military capabilities, and posed a serious threat to both the mountainous communities and the agricultural dynasties in the Central Plains.

Amidst conflict and confrontation, pastoral societies and agricultural societies followed distinctly different paths of development: one emphasizing mobility evolved into nomadic states on the grasslands, while the other emphasizing stability transformed into agricultural states. The former pastoral communities in the mountainous regions found themselves caught between two powerful adversaries. They lacked the mobility and combat prowess of the grassland groups while also lacking the manpower and resources of the agricultural societies. Gradually, they were assimilated by the stronger entities on both sides.

For instance, the historical Shanrong (山戎), referred to as the upper layer of the Xiajiadian Culture by archaeologists, were followed by the northeastern groups that conflicted with the lower layer of the Xiajiadian Culture, which was in turn in conflict with the Shimao Culture. The Shanrong established a regime that combined both farming and animal husbandry. During the Eastern Zhou Period, the powerful Shanrong inflicted significant damage on northern vassal states like Yan (燕) until Duke Huan of Qi (齐桓公).

During the Spring and Autumn Period, Duke Huan of Qi rallied the Eastern Zhou states to "honor the king and repel the barbarians" (尊王攘夷), unifying the vassal states and expelling the Shanrong to the north. The Shanrong likely perished due to pressure from both the central plains kingdoms and the steppe nomadic powers, paving the way for the rise of a nomadic power in that region—the Donghu (东胡).

Another mountainous regime closer to the Central Plains was Guzhu (孤竹), which occupied the areas both north and south of the Yanshan Mountains and in the Taihang Mountains. It was a key target of attacks by the Central Plains dynasties. After Duke Huan of Qi's campaign against Guzhu, he got lost on his way back but found his way using an old horse's memory, leading to the origin of the idiom "an old horse knows the way" (老马识途).

Genetic research also reflects the dilemma of ethnic groups in mountainous regions. For instance, in the Zhukaigou site near the border of Inner Mongolia and Shaanxi, dating back from over 4,000 years ago to around 3,500 years ago. By conducting DNA identification on ancient human teeth unearthed from this site, it was found that the major genetic profile belonged to populations from Eastern Asia, particularly those from Northern and Central China. During that era, the population in the Zhukaigou region likely engaged in pastoral economies, and genetically, they exhibited similarities with the populations in the southern Central Plains region.

Not far to the east of the Zhukaigou site, there is another archaeological site, likely from the later period of the Eastern Zhou Dynasty, around the Warring States Period. From a genetic analysis perspective, the genetic profile of the later population at this site continued the genetic profile of the Zhukaigou population. However, from the local tombs, two distinct burial styles emerged: in one type of tomb, the deceased were oriented with their heads to the east, while in the other type, the heads were oriented to the north. Archaeologists speculate that the former burial style might belong to nomadic groups, while the latter could be associated with Central Plains ethnic groups. This indicates that both the agrarian and nomadic cultural influences from the north and south exerted their impact on this intermediate region.

So, how did the concept of "中国" in the sense of Chinese culture come into being in people's minds?

It began with neighboring regions; hence, China exists.

In order to defend against various ethnic groups different from their own culture, especially powerful nomadic groups from the north and west, the concept of "中国" developed among the Chinese ethnic group. China had fundamental economic and political differences with its neighboring regions, resulting in cultural and civilizational distinctions. If we extend our perspective and broaden it, we will discover that the emergence of Chinese culture is both the outcome of the development of the various ethnic groups on the Chinese land and the result of transformative forces coming from the northwest region and even the entire Eurasian steppe belt.

This is the butterfly effect on the Eurasian continent, where the landmass was closely interconnected, and no region existed as an isolated island.

5.4 Ethnic Expansion: From Biological Genes to Cultural Memes

From over 4,000 years ago to over 2,000 years ago, in the "brief" span of 2,000 years, the entire Eurasian continent transformed from picturesque wilderness into crowded bustling cities. Whether in the fertile plains and valleys suitable for agriculture, the grasslands suited for nomadic herding, or the forests suited for hunting and fishing, the populations of each ethnic group underwent significant growth. Throughout the continent, different ethnic groups jostled for space, leading to continuous interaction and friction.

Looking at this period from a genetic perspective, our ancestors expanded their territorial influence as ethnic groups, consequently extending their genetic distribution. This process can be clearly divided into three successive modes.

The first genetic expansion mode is what the second chapter of this book calls the "filling-in-the-blank mode." When there were still many uninhabited virgin lands on the continent, ethnic expansion primarily involved occupying the vacant spaces in the natural environment. Ethnic groups engaged in farming or herding migrated to new territories uninhabited by other humans, transforming those areas into their homelands.

Of course, saying these virgin lands were completely devoid of human presence is not entirely accurate. Hunter-gatherer populations once spanned the entire continent, but their population density was low, and their group sizes were small. Therefore, when farming and herding ethnic groups encroached upon their territories, the small numbers of hunter-gatherers were either eliminated, assimilated, or retreated deeper into the forests and mountains.

In the "filling-in-the-blank mode," the genetic makeup of a new territorial population was consistent with their homeland ethnic group, at least in the short term. There were distinct genetic differences between different ethnic groups, making it easy for molecular biologists to trace the origins of these groups. Due to their differing origins, ethnic groups inherited their respective cultures during expansion, resulting in significant cultural differences between them. The southward migration of people from Karamay to Lop Nur could be considered an example of this mode of expansion.

The second genetic expansion mode is the "group replacement mode." At this point, uninhabited virgin lands were difficult to find; every region was inhabited by various ethnic groups with varying sizes and distinct cultures. These groups remained relatively unfamiliar with each other, with limited interactions and significant cultural and even linguistic differences, causing them to view each other as incomprehensible "others." Under pressures from environmental changes or purely for the sake of securing better living conditions, powerful ethnic groups would encroach upon the territories of weaker ones. Many smaller ethnic groups either perished or were driven to other regions, leading to a replacement of the population in that space.

In the "group replacement mode," the genetic makeup of the territorial population was replaced, transitioning from the original indigenous genetic makeup to that of the new inhabitants. Perhaps a very small number of members from weaker ethnic groups were assimilated by the dominant ethnic group, but due to the vast difference in population size and social status, these few members had few or no descendants, thereby not contributing new genetic traits to the gene pool of the dominant ethnic group. The process of the Okunev people displacing the Qiemu'erqieke people and the Karasuks displacing the Altai tribes broadly falls into this mode.

The third model of gene expansion is the "upper-level replacement mode." In this scenario, various ethnic groups have substantial populations and some level of understanding between them, including language, culture, and economic exchanges. Wars between ethnic groups are conducted using specialized military forces. Ultimately, the ruling power of a certain ethnic group is overthrown. The previous upper-level royalty and nobility are replaced by new upper-level royalty and nobility, while the common people remain in place, subject to the rule of the new governing class.

In the "upper-level replacement mode," the genetic changes within a region can be quite complex. The genetic makeup of the broader population at the lower levels remains relatively unchanged, while the genetic makeup of the upper-level rulers may undergo replacement, differing from that of the lower-level population. It's also possible that the new rulers are genetically similar to the previous rulers, in which case the change in regime might not bring about genetic changes, whether among the upper-level royalty, nobility, or lower-level populace. In such cases, the changes occur in cultural genes rather than biological genes.

The concept of cultural genes can be traced back half a century to the renowned British science writer Richard Dawkins [b. 1941], who introduced the term "meme" in his book "The Selfish Gene" to describe "cultural replicators." Dawkins proposed that the process by which an idea or behavior in a culture is imitated, spread, and becomes popular among a population is quite analogous to the biological process of genetic inheritance and transmission. The competition between different memes or "cultural genes" resembles, in principle, the competition between biological genes. As human societies developed diverse cultures, competition between cultural genes also emerged.

In the "upper-level replacement mode," the replacement of biological genes might be relatively limited, but the replacement of cultural genes can be more significant and impactful on ethnic groups. Thus, as ancient societies entered an era of intergroup competition with large populations, competition between cultural genes became predominant. The divide between certain ethnic groups and ruling powers, as opposed to other groups and powers, might not be due to biological genes but rather cultural genes.

Taking the expansion of ethnic groups in the Central Plains region as an example, as agriculture began to support larger populations, agrarian communities expanded into open regions with cultivation potential. These groups brought with them crops like foxtail millet and proso millet from the north, rice from the south, or wheat and barley from the west, transforming new landscapes into farmland. During this phase, genetic flow can be described as a "filling-in-the-blank mode." When various cultural groups emerged and competed for valuable resources and environments, the "group replacement mode" came into play. For instance, dominant cultural groups from the Central Plains colonized weaker ethnic groups in southern or northern mountainous regions.

Moving forward, historical events like the Zhou Dynasty replacing the Shang Dynasty exemplify characteristics of the "upper-level replacement mode." The Zhou Dynasty, a smaller state, overthrew the Shang Dynasty, a larger state. The ruling elite of the Shang Dynasty were replaced by those of the Zhou Dynasty, while the commoners of the Shang Dynasty became the commoners of the Zhou Dynasty, continuing their agricultural or pastoral lifestyles.

From a genetic standpoint, the upper-level replacement of the Zhou Dynasty might have introduced some genetic elements from the northwest to the Central Plains region. However, this exchange likely began before the Zhou Dynasty's ascension over the Shang Dynasty. Genetic changes in the Central Plains ethnic groups were minor. What truly changed was the cultural gene pool. The Zhou Dynasty's ceremonial culture gradually replaced the Shang Dynasty's culture that included beliefs in spirits and gods. This shift reinforced the cultural identity of the Central Plains and broader regions as "Chinese" culture.

During the Spring and Autumn Period, Confucius' commentary on the statements of Guan Zhong vividly depicted the cultural identity of the noble class at the time. According to the Analects (论语), Zigong (子贡) said to Confucius that Guan Zhong couldn't be considered a virtuous man. Even though Duke Huan of Qi killed Guan Zhong's previous lord, Duke Hui, Guan Zhong did not die in mourning but instead aided Duke Huan. Confucius replied that Guan Zhong had helped Duke Huan establish dominance over other feudal lords, rectify the order of the realm, and the people continued to benefit from Guan Zhong's contributions. "But for Guan Zhong, we might be wandering about with unbound hair, and the lappets of our coats flying behind us" (微管仲, 吾其被发左衽矣).

Confucius' focus was on whether their hair was tied or untied and whether their clothing opened to the left or the right. This represented the cultural identity of the ethnic group. During that era, the Central Plains dynasties with shared cultural genes were embarking on "cultural gene expansion" in all directions, leveraging their abundant grain production and large population. The trend of genetic and cultural expansion, previously blowing from the west to the east, was about to reverse significantly, as the second half of the West's dominance in the Central Plains commenced.

The Column: Milk has Shaped the Course of Human History

Since the dawn of human ability to tame large mammals—be they camels, horses, cows, or sheep—humanity has partaken in an exquisite banquet of nourishment: animal milk, a treasure trove of premium protein. Yet, the enigma lay in effectively assimilating its nutritive gifts. This challenge arose from the imperative of mammalian metabolism, where the conquest of milk's bounty demanded the aid of lactase enzymes. These biological catalysts are requisite for the deconstruction and digestion of lactose, the protein inherent to milk's constitution. However, most mammals, including humans, possess this enzymatic wizardry only in infancy, the potency dissipating as maturity unfurls its wings. Humans were no strangers to this biological script.

Enter the fortuitous genetic mutiny, bestowing upon certain groups the extraordinary lactase gene, enabling them to wield its power even into adulthood. This enzymatic clemency, offering the ability to digest animal milk, emerged approximately 9,000 years ago, likely originating within the expansive embrace of Eurasia and Africa's western precincts. Among the modern populace, this genetic legacy thrives predominantly within Europeans and Africans, while its presence dwindles among Asians and Native Americans. Notably, a vast majority—around 85% to 100%—of modern individuals from East and Southeast Asia are bereft of this genetic blessing, a stark contrast to the underwhelming 10% scarcity within North European populations.

The ancient custom of indulging in animal milk may have its roots tracing back millennia. Archaeological whispers in the form of lipid residues adhering to the fragments of ancient pottery herald Asia's western realms as potential cradles of this culinary legacy. This antiquarian narrative synchronously waltzes with the advent of the lactase gene, propelling a hypothesis that their debut on the world stage was a synchronous affair. Could it be that humanity engaged in numerous *pas de deux* [step of two] with animal milk throughout history, yet fortune's embrace evaded them until an epiphanic moment of harmonious digestion emerged? Thereafter, animal milk gracefully pirouetted into the realm of daily sustenance for these custodians and their lineage.

Tangible evidence of this gastronomic ballet surfaces through the prism of ancestral dental calculus. Peering into the lives of those who dwelled some 5,300 years ago within the heartland of Mongolia and the shadow of the Altai Mountains, scientists reveal a tableau of cow and sheep milk, penned by the proteins of dental remnants. The chronicle etched on these ancient dental tablets coalesces with the tale of the lactase gene, implying a symphonic inception. Perhaps, across a span of 5,000 years, the waltz of milk and the spread of domesticated creatures twined their fates, embracing the dispersal of the lactase gene in their pas de deux. Archaeological murmurs intimate that a cultural exodus flowed eastward 5,000 years ago from the western realms of Eurasia.

China's earliest footprint in the annals of milk consumption unfurls from the Xiaohe Cemetery in Xinjiang. Here, the custodians of history lay enshrouded in the faint golden embrace of milk remnants, a mosaic of custodial care painted upon their necks and chests. Within burial baskets, granulated tales echo these imprints, beckoning inquiry. These relics, deciphered through the lens of analysis, unveil themselves as predominantly bovine brews, with a mere whisper of ovine essence.

As the molecular veil lifts further, protein analysis distinguishes these granular relics from their lumpen counterparts. The granules, a testament to whole milk's embrace, bear witness to life's creamy continuum. Conversely, the lumps—replete with Casein but a stranger to Whey—stand as paragons of cheese, artifacts of an ancient curd crafted from bovine nectar through the alchemy of Lactobacilli, now christened "kefir cheese." With this revelation, history's pendulum swings back 3,600 years, bringing the legacy of Kefir cheese into the light of antiquity, an honor bestowed upon China's cradle.

In an analogous chronicle, whispers of mare's milk beckon from ancient Mongolian dental scrolls, scribed some 3,200 years ago. A symphony of consumption converges with the ascension of equine mastery within the Eastern Eurasian steppe, illuminating a metamorphosis upon these pastoral canvases. This tableau heralds a transition, as settled societies navigate the labyrinthine corridors from "agriculture-livestock" to the transcendence of nomadic winds. By the empire's zenith, the script of mare's milk weaves seamlessly into the daily fabric.

In the theater of needs, dairy's bounty—culled from equine, bovine, and ovine sources—cascades with opulence, outshining their meaty counterparts in caloric opulence, not to mention their nutrient ballet. A tapestry of domestication unfolds, where horses, cows, and sheep emerge as refined muses. The touch of human refinement, a symphony of genetic refinement, guides these creatures along the path to docility, culinary opulence, and prodigious lactation. The chasm between untamed aurochs and their domesticated kin—manifest in both size and temperament—whispers to humanity's indelible mark upon the annals of bovine evolution.

As the pages of history unfurl, the enigma lingers: did the lactase gene's steppe-bound migration forge a steely mettle within its beneficiaries, rendering them potent orchestrators of regional destiny? An inquiry ripe with nuances, awaiting the scholar's quill.

However, southward of these pastures, a different refrain danced. Amidst the annals of Huaxia—ancient China—a dearth of the lactase gene prevailed.

Even as yaks, sheep, and the equestrian maestros graced the verdant plains, the artisans of this land, the farmers, rarely partook in milk's velvety embrace. This culinary trove, esteemed in the fertile crescents and sunlit steppes, remains a stranger to this land, where these creatures are revered for their sustenance, their labor, and their hides.

Had ancient East Asia birthed the lactase gene's ballet, while the West danced a different waltz, history's tune would have spun a different tale.

CHAPTER 6

Reversing Horizons: The Han and Xiongnu Unleash a Westward Surge

As the coffin lid was raised by the collaborative effort of several archaeologists, the intertwined remains of a man and a woman, resting alongside burial artifacts untouched by the passage of time, faithfully preserved the moment of their interment. Amidst the fabrics adorning the tomb's occupant, a resplendent brocade armband emerged like a vibrant gem amidst the muted shroud of dust in the tomb chamber, instantly drawing the gaze of the archaeologists. As this brocade was gently unfurled, amidst a symphony of colors—blue, white, red, yellow, and green—were discernible Han characters, including "国" (country), "东方" (eastern), and "五星" (five stars), etched in blue upon a white canvas. When the entirety of the brocade was unveiled, the assemblage of Han characters coalesced into a sentence that prompted those present to involuntarily exclaim, "Five stars rising in the East, a harbinger of fortune for the Middle Kingdom! (五星出东方利中国)"

This enthralling archaeological expedition unfolded in the Niya ruins of Hotan, Xinjiang, during the crisp month of October in 1995. Remarkably, a day prior to the unsealing of the coffin lid, an archaeologist hailing from Peking University, bearing the name Qi Dongfang (born in 1956) (齐东方), serendipitously arrived at the Niya ruins bearing aloft a five-starred red flag (五星红旗). The alignment of the phrase "Five stars rising in the East, a harbinger of fortune for the Middle Kingdom" on this magnificent brocade embroidered a tale of enchanting synchronicity between archaeologist Qi Dongfang and the five-starred red emblem—a narrative that would subsequently become a cherished anecdote within the realms of archaeological discourse.

In truth, the extraordinary narrative surrounding this brocade remains incomplete, for the ethnic evolution unveiled by the characters adorning the fabric is equally mesmerizing.

6.1 Chadoma and the Southern Qiang Hidden in Brocade

The coincidences between the inscriptions on brocade and the Qi Dongfang, the Five-Star Red Flag, and China have sparked discussions among people, but from a scholarly perspective, these are merely coincidences. This piece of brocade was woven in the imperial workshop of the Western Han Dynasty, and as mentioned in the previous chapter of this book, the ancient Chinese interpretations of "China" varied over different periods. When the Western Zhou people said, "Here in China I Dwell" (宅兹中国), they may have referred to the Lo region where they conducted sacrifices, corresponding to the celestial capital in the heavens.

During the Western Han Dynasty, "中国" (the Middle Kingdom) essentially referred to the Lo region mentioned by the Eastern Zhou people and the surrounding Central Plains area. It was a geographical concept rather than a national name. In the Western Han Period, "五星" (Five Stars) referred to the five major planets visible to the naked eye in the night sky: Jupiter, Saturn, Mercury, Mars, and Venus. After the rise of Daoist Yin-Yang and Five Elements theories[1] in the Eastern Han, these five stars were renamed as Jupiter (木星), Saturn (土星), Mars (火星), Venus (金星), and Mercury (水星), the Chinese names that are still in use today.

The phrase "五星出东方利中国" (Five stars rising in the East, a harbinger of fortune for the Middle Kingdom) referred to a celestial phenomenon where all five planets converged in the eastern sky, signaling great prosperity for the Chinese region. Conversely, if they converged in the western sky, it was considered beneficial for foreign powers.

---

1 The theory of five elements explains how *Qi* (all the vital substances) cycles through various stages of transformation. As yin and yang continuously adjust to one another and transform into one another in a never-ending dance of harmonization, they tend to do so in a predictable pattern.

In both Chinese philosophy, these stages of yin-yang transformation are referred to as the "five phases of transformation (*wu xing*)." The five elements are listed below:

Water (水) is associated with the potential of new life hidden in the dark ground beneath the snows of winter.

Wood (木) is associated with the exuberance of new growth as it shoots up from the earth in the spring.

Fire (火) is associated with the process of maturation that takes place under the warmth of the summer sun.

Earth (土) is associated with ripening of grains in the yellow fields of late summer.

Metal (金) is associated with the harvest of autumn and the storage of seed for next year's planting and a new cycle.

—Translator's note.

However, the phrase "五星出东方利中国" is actually incomplete. A fragment of brocade was also unearthed in the tomb, which is the same fabric as the "five-star" brocade, with identical patterns. This brocade fragment bears three characters, "讨南羌" (put down South Qiang). Therefore, the complete inscription on the brocade should read, "五星出东方，利中国讨南羌" (Five stars rising in the East, a harbinger of fortune for the Middle Kingdom in subduing the Southern Qiang).

At that time, the Niya site was part of the Chadoma kingdom, and a pottery jar inscribed with the character "王" (king) in the tomb suggests that the tomb's occupant was likely the king of Chadoma. So, why did the distant Western Han Dynasty have a connection with Chadoma? And where is the "South Qiang?"

Let's first explore the history of Chadoma through its genetics.

The Niya site was discovered in 1901 by the British explorer Marc Aurel Stein, (1862–1943) who removed a large number of artifacts from the site, causing a sensation in the European academic community. The Niya site is sometimes referred to as the "Pompeii of the East." In 1995, a joint Chinese-Japanese archaeological team conducted further excavations at the Niya site. Among the artifacts unearthed were Western Asian-style glassware, Greek-style artworks, Gandharan[2] decorative patterns, ancient Indian cotton textiles, Central Plains silk, satin, lacquerware, bronze mirrors, and pieces of paper. This site represents a rare convergence point of four ancient civilizations: ancient China, ancient India, ancient Greece, and ancient Persia.

Scientists conducted mtDNA analysis on a male adult's femur bone from the Niya site and found that his genetic profile belonged to the Western Eurasian genetic type, closest to West Asian and Iranian populations, with no genetic connection to East Asian populations. Comparing the ancient human DNA from other sites in Xinjiang, scientists found that the DNA of a population from a site in Yutian, which is in the same region as Chadoma, was similar to the Niya site population. Yutian and Chadoma were both located in the southern edge of the Tarim Basin and, historically, Yutian even annexed Chadoma. Therefore, the genetic similarity between these two populations and West Asian and Iranian populations suggests some historical population migrations.

---

[2] Gandhara was an ancient Indo-Aryan civilization centered in the present-day Northwest Pakistan and including parts of Northeast Afghanistan, roughly in the northwestern part of the Indian subcontinent. The core of the region of Gandhara was the Peshawar Valley and Swat Valley, though the cultural influence of "Greater Gandhara" extended across the Indus river to the Taxila region in Potohar Plateau and westwards into the Kabul Valley in Afghanistan, and northwards up to the Karakoram range.
—Translator's note.

According to radiocarbon dating, the Niya site dates back to 1,500–2,200 years ago, extending from the Western Han Period to the period of the Northern and Southern Dynasties. Meanwhile, the Yutian site dates back to 1,700–2,200 years ago, from the Western Han Period to the Three Kingdoms and Two Jin Periods. Scholars speculate that before the emergence of Chadoma, there was likely a population with West Asian and Iranian genes that crossed the Pamir Plateau, entered the southern edge of the Tarim Basin, lived in oases between the Kunlun Mountains and the Taklamakan Desert, and possibly expanded eastward. During the Western Han Dynasty, Emperor Wu of Han sent Zhang Qian on a mission to the Western Regions, breaking down barriers between the Han Dynasty and various Western Region states, including Chadoma, and thus, the Han Dynasty began to exert influence on these states.

In terms of genetics, the Chadoma population had a Western genetic profile, distinct from the East Asian genetic profile of the Han Dynasty. However, this difference did not hinder friendly relations between Chadoma and the Han Dynasty. The royal brocade with the inscription "五星出东方，利中国讨南羌" was likely a gift from the Han Dynasty to the king of Chadoma, regarded as a mysterious object with the power to suppress enemies and thus buried with the king upon his death.

So, where is the "Southern Qiang" to be conquered?

The Southern Qiang, also known as the Southern Mountain Qiang, refers to the Kunlun Mountains. After Zhang Qian returned to the Han Dynasty from his mission to the Western Regions, he reported to Emperor Wu of Han that the source of the Yellow River was in the Kingdom of Khotan (于阗国). According to ancient legends, the Kunlun Mountains should be further west. Based on the mountains from which the river originated in the Kingdom of Khotan, which also happened to produce jade, it closely matched the description of the Kunlun Mountains in ancient texts. Therefore, Emperor Wu of Han proclaimed that the Southern Mountains of the Kingdom of Khotan were the Kunlun Mountains.

Now we know that the Yellow River originates from the Bayan Har Mountains, not the Kunlun Mountains, and the legendary Kunlun Mountains mentioned in ancient texts are elusive and cannot be pinpointed to an actual mountain range. At that time, Zhang Qian and Emperor Wu of Han did not have the scientific knowledge and literacy that we have today, so their judgment was understandably flawed. However, the name Kunlun Mountains given by Emperor Wu of Han has persisted to this day and refers to a prominent mountain range along the southern edge of the Tarim Basin. At that time, oasis states in the Western Regions commonly referred to it as the Southern Mountains.

The Southern Qiang were a nomadic group on the Kunlun Mountains. In terms of their ethnic origin, the Southern Qiang likely originated from ancient Qiang tribes and incorporated members from neighboring tribes. The

Southern Qiang initially lived under the Han Dynasty's Jimi system (羁縻) but later attempted to cross the Huangshui River in search of suitable land for settlement, leading to a rebellion. They may have had the support of the Han Dynasty's archenemies, the Xiongnu. In 61 BC, General Zhao Chongguo, at the age of 70, led an expedition to suppress the Qiang rebellion. Before departing, Emperor Xuan of Han granted a decree stating, "Now the five stars rise in the east, bringing great advantage to China; the savages and barbarians shall be utterly defeated (今五星出东方，中国大利，蛮夷大败)." Zhao Chongguo quickly quelled the rebellion and then actively engaged in farming in the region, ensuring long term stability.

Therefore, the embroidered silk with the phrase "五星出东方，利中国讨南羌" was likely a "propaganda piece" produced after this victory, and it was probably presented as a gift to the oasis states near the Kunlun Mountains, with Chadoma receiving one.

The people of Chadoma belong to the genetic profile of the Western Eurasian continent, so are the Southern Qiang people native to the genetic profile of the Eastern Eurasian continent? Not necessarily. Ancient Qiang people and their direct descendants primarily belong to the genetic profile of the Eastern Eurasian continent. However, the Southern Qiang people not only inherited the genetic profile of the ancient Qiang people but may have also incorporated genes from another significant group, the Lesser Yuezhi (小月氏) people. The Lesser Yuezhi people have a genetic profile of the Western Eurasian continent. Therefore, historically, the Southern Qiang may have been a mixture of Eastern and Western influences, with their ethnic and genetic backgrounds merging.

6.2 The Xiongnu and Their Grassland Rivals

If there were the Lesser Yuezhi, then there must have been the Great Yuezhi (大月氏) as well. And when we talk about the Yuezhi, we inevitably introduce the arch-enemy of the Qin and Han Dynasties—the Xiongnu. To understand the evolution of ethnic groups in the northwest region, the Xiongnu are an essential group that cannot be ignored.

Historically, the Xiongnu emerged on the stage of history during the Warring States Period and began to threaten the northern powers of the Seven Warring States, namely Qin, Zhao, and Yan. Their influence continued until the Eastern Han Dynasty when they were defeated by the Han Dynasty, along with the Xianbei (鲜卑) and other allies. The Xiongnu engaged in a centuries-long struggle with the Han Dynasty, centered around the Central Plains. However, many details about this significant nomadic power remain unclear, particularly the origins of the Xiongnu as a people.

The challenge lies in the fact that the Xiongnu were a large tribal alliance in the Eastern Eurasian grasslands, encompassing various tribal backgrounds from a vast region. It is highly likely that they incorporated genetic influences from Siberia, Eastern Asia, Central Asia, Western Asia, and even Europe. Therefore, in terms of their expansion pattern, the Xiongnu followed a typical "upper-level replacement mode." They likely originated from a core tribe and, through a process of continuous conquests, absorbed various tribal groups into their rule. The original upper-class nobility of these tribes either got replaced or submitted to the Xiongnu Chanyu (单于), becoming agents of the Chanyu in governing their respective tribes.

Since the Xiongnu's genetic sources were diverse and complex, it is challenging to determine the origin of the Chanyu clans, which constituted the core of the Xiongnu ruling elite.

Molecular biologists once conducted genetic testing on over 60 skeletal remains from a Xiongnu-era burial site within Mongolia, dating back approximately 2000 years ago, during the height of the Xiongnu's dominance over the Mongolian Plateau. The analysis revealed that nearly 90% of the individuals in the burial site had genetic profiles typical of the Eastern Eurasian mainland, while about 10% had genetic profiles more characteristic of the Western Eurasian mainland. Scholars also conducted genetic tests on three skeletal remains from another Xiongnu noble burial site in Northeastern Mongolia, revealing one individual with a western genetic profile and two with eastern profiles.

Therefore, based on genetic analysis of the Xiongnu people from burial sites, the majority of the Xiongnu population likely belonged to indigenous ethnic groups of the Eastern Eurasian mainland, primarily from various tribes in the Eastern Eurasian grasslands. However, at least 2,000 years ago, a portion of the Xiongnu population may have had origins in the Western Eurasian mainland or its descendants. While the genetic proportions are established, it remains uncertain whether the Chanyu clans of the Xiongnu belonged to the majority of Eastern Eurasian mainland ethnic groups or the minority of Western Eurasian mainland ethnic groups.

Linguistic analysis faces a similar dilemma. For decades, linguists have attempted to determine the language spoken by the Xiongnu by comparing the phonetic transliterations of certain Xiongnu-specific terms in the records of the Qin and Han Dynasties with the languages of various ethnic groups. Given that the Xiongnu were a large tribal alliance, it is certain that different languages were spoken among the Xiongnu themselves. However, the specialized vocabulary recorded in Chinese texts is likely to have been used by the upper echelons of the Xiongnu, representing the language of the Chanyu clan. Linguists have long debated whether the language spoken by the Xiongnu belongs to the Turkic or Mongolic language systems, with no definitive conclusion.

Both genetics and linguistics have encountered challenges in tracing the origins of the Xiongnu Chanyu clans, highlighting the limitations of these disciplines when researching the origins of ethnic groups.

From a molecular biology perspective, scholars compare the DNA of ancient individuals with the DNA of modern people living in Europe, Africa, Asia, and the Americas to determine the ancestral origins of ancient people. However, modern humans are descendants of ancient people, not the other way around. Using the genes of modern people to determine the ethnic categories of ancient people is logically flawed. For example, the ancestors of modern Europeans might not have lived in Europe but rather in Western Asia, so the genetic profiles of ancient European ancestors might be more similar to certain Western Asian ethnic groups than to modern Europeans. When we find ancient DNA in Europe that is closer to modern Western Asian populations and farther from modern European populations, we must recognize that these ancient people may well have been the ancestors of Europeans.

Similarly, for linguistics, comparing the language of the Xiongnu to modern Mongolic and Turkic languages is an inversion of priorities. The reality is that the language of the Xiongnu may have influenced the formation of modern Mongolic and Turkic languages. Many Xiongnu vocabulary and grammatical features were inherited and used by later peoples, including the Mongols, and have survived to the present day. Therefore, later researchers can find evidence of linguistic similarities between the language of the Xiongnu and modern Mongolic and Turkic languages, but it cannot be said that the Xiongnu spoke a so-called ancient Mongolic or ancient Turkic language.

So, what can genes and language tell us? They can inform us about the relationships and distances between various modern ethnic groups. Furthermore, genes can tell us about the approximate time when two populations initially separated through mutation rates. It's essential to be aware of the limitations of genetic and linguistic research.

Returning to the discussion of the origins of the Xiongnu Chanyu clan, if we can discover a significant number of Xiongnu burial sites and determine which ones belong to the Xiongnu Chanyu clan and which ones belong to ordinary Xiongnu individuals, and then analyze their DNA separately, it should provide some assistance in exploring the origins of the Xiongnu Chanyu clan. In addition to genetic and linguistic research, various other sources of information must be utilized for analysis, including archaeological findings and records in ancient texts.

The "Treatise on the Xiongnu" (匈奴列传) in the Records of the Grand Historian (*Shi Ji*) states: "At that time, the Donghu were strong, and the Yuezhi were flourishing (当是之时，东胡强而月氏盛)." During the late Warring States Period to the Qin Dynasty, the northern grasslands of China witnessed the dominance of three powers: the Yuezhi to the west, the Xiongnu in the middle, and the Donghu to the east. The early influence of the Xiongnu was restricted by the two powerful neighbors to their east and west. Based on this grassland landscape, it is possible that the Xiongnu Chanyu clan was a local group from the Mongolian Plateau, or they may have migrated south from the region around Lake Baikal in the far north, rather than being a steppe group from Central or Western Asia.

Especially noteworthy are the Yuezhi, who were located to the west of the Xiongnu. In terms of ethnicity and language, the Yuezhi are likely descendants of the Karasuk people, so genetically, they may have more Western Eurasian components. The Yuezhi ethnic group likely spoke an ancient language known as the Tocharian (吐火罗) languages, which is why they and related groups are often referred to as "Tocharian people." Tocharian activities within China can be traced back to the Qiemu'erqieke people and the groups that migrated southward to the Lop Nur region, known as the "Xiaohe Princess."

The Yuezhi were active to the west of the early Xiongnu, but there is some scholarly debate about their specific territory. Chinese historical records mention that the Yuezhi and another group, the Wusun, were "together in Dunhuang" (共在敦煌间) and "both in Qilian and Dunhuang" (俱在祁连、敦煌间). If we use present-day geographical names, Qilian Mountain is located at the border of Qinghai and Gansu, marking the southern boundary of the Hexi Corridor, while Dunhuang is in the western part of Gansu, at the far western end of the Hexi Corridor. Therefore, one would expect the Yuezhi to be active in present-day Gansu's Hexi Corridor and the neighboring northern regions of the Mongolian Plateau.

However, genetic research challenges this view. Genetic studies of ancient human DNA from sites in Gansu, Ningxia, Shaanxi, Shanxi, and other ancient sites dating to the period of Yuezhi activity indicate that these regions primarily had Eastern Eurasian genetic components and lacked Western Eurasian genetic components. During the same period, Xinjiang in the west hosted both Eastern and Western Eurasian genetic components, suggesting that Western groups may have expanded eastward into Xinjiang but might not have reached the Hexi Corridor in Gansu.

So, how can we explain the historical records mentioning the Yuezhi being "together in Dunhuang" and "both in Qilian and Dunhuang?" Just as we discussed the variation in the name Kunlun Mountains earlier, the term "Qilian Mountains" (祁连山) in ancient times did not necessarily refer to the modern Qilian Mountains. According to linguists' research, whether it's the Kunlun

Mountains, Qilian Mountains, or today's Tianshan Mountains in Xinjiang, their names all actually come from the Tocharian word for "sacred sky." These names in Chinese historical records were likely based on the Tocharian people's appellations for these mountain ranges. Therefore, "Qilian and Dunhuang" likely refers to locations within present-day Xinjiang, possibly the Tianshan or Kunlun Mountains, and could even include the Altai Mountains.

In conclusion, the areas of Yuezhi activity were in the northern grasslands of present-day Xinjiang and further north, revealing new insights brought about by genetic research.

The Yuezhi were once very powerful to the extent that the Xiongnu had to send political hostages to the Yuezhi. In the Records of the Grand Historian, it is described: "The old Chanyu had a crown prince named Modu. Later, he fell in love with a lady named Yanzhi, who gave birth to a younger son. The Chanyu wished to depose Modu and make the younger son the successor, so he sent Modu as a hostage to the Yuezhi" (单于有太子名冒顿。后有所爱阏氏，生少子，而单于欲废冒顿而立少子，乃使冒顿质于月氏). Modu (冒顿) spent some time as a hostage among the Yuezhi before becoming the Chanyu of the Xiongnu. This indicates that at the time, the Yuezhi were more powerful than the Xiongnu on the grasslands.

However, when Modu killed the old Chanyu who wanted to kill him, and became the new Chanyu of the Xiongnu, the Xiongnu entered a triumphant era. They defeated the Yuezhi to the west, killed the Yuezhi leader, and used his skull as a drinking vessel. The Yuezhi were defeated by the Xiongnu and moved far away, passing through Wuan, heading west to conquer Daxia and establishing their capital north of the Oxus River. The remaining smaller groups who couldn't leave sought refuge in the Southern Qiang territory.

The Yuezhi's westward migration was like a domino effect, except this time, the dominoes fell from east to west. The Yuezhi first occupied the Ili River (伊犁) Valley, which was previously held by the Wusun people. Perhaps due to the swift advance of the Xiongnu, the Yuezhi had to continue their westward migration, crossing the Amu Darya River, capturing Daxia (大夏), and making the city of Lanzhi their capital.

Previously, based on records that mentioned the Yuezhi and Wusun being active "between the Qilian Mountains and Dunhuang," it was thought that the Wusun had also been active in the Hexi Corridor and later occupied the Ili River Valley after the Xiongnu drove out the Yuezhi. However, if genetic research is correct and the Yuezhi and Wusun's influence was within Xinjiang, then the Wusun may have originally been active in the Ili River Valley, south of the powerful Yuezhi. In this scenario, when the Yuezhi attacked from the northeast, the

Wusun were the first to bear the brunt of the attack. The remnants of the Wusun fled east to take refuge in the Turpan Basin, which makes sense geographically.

Under the Xiongnu's onslaught, the Yuezhi split into two groups. The main force moved west, occupied lands in Central Asia, and later established the Kushan Empire. Chinese historical records still referred to them as the "Great Yuezhi." The smaller groups that didn't migrate westward fled south from the Ili River Valley and merged with local populations, forming the Southern Qiang, the target mentioned in the saying, "Five stars rising in the East, being a propitious sign for the Middle Kingdom to put down South Qiang."

Another powerful entity among the northern grassland's three major players, the Donghu, also suffered a devastating blow from the rising Xiongnu. The Records of the Grand Historian vividly describe how the Donghu repeatedly extorted the Xiongnu, demanding prized horses, beautiful women, and land. After enduring several demands, Modu finally led the Xiongnu to crush the complacent Donghu. After a considerable period, the Xianbei and Wuhuan emerged in the Eastern Mongolian Plateau, with the Xianbei in the north and the Wuhuan in the south. Both groups threatened the Central Plains during the late Eastern Han Dynasty. It was suggested that after the Donghu disintegrated, some of them sought refuge in the Xianbei Mountains to the north, while others sought refuge in the Wuhuan Mountains to the south, and both the Xianbei and Wuhuan were descendants of the Donghu.

If the grasslands were still in a "group replacement mode" during that time, meaning that a group migrating would largely maintain their genetic characteristics, then the claim that the Donghu were ancestors of the Xianbei and Wuhuan wouldn't be unreasonable. However, during the era when the three major northern grassland powers—the Xiongnu, Yuezhi, and Donghu—coexisted, and even before that, the grasslands had already shifted into an "upper-level replacement mode." When the Xiongnu defeated the Donghu, they likely only expelled a small upper-class faction, while the majority of the population likely remained in place. They underwent a transformation, changing from Donghu to Xiongnu. This was a common scenario on the ancient grasslands, where numerous tribes would change allegiances with the rise of a dominant power, adopting the name and identity of that power, as discussed earlier in terms of cultural genes and cultural identity.

Similarly, when the Yuezhi migrated westward, it was only a portion of their population that moved. Even though the Yuezhi were a grassland power, they still had enough military strength to defeat some Central Asian tribes and drive away their upper-class elites, taking control of their territory and population. Take the capital city of Balkh (蓝氏城) in Daxia as an example. This name is a transliteration from Chinese historical records and originally meant "Alexandria

City," a city established by Hellenized tribes that remained in the region after the conquests of Alexander the Great. Clearly, these Hellenized tribes constituted the upper class of Daxia, ruling over the local indigenous population. The Yuezhi merely replaced the ruling elite once again.

6.3 The Eastern Wind Disperses the Dreams of Loulan (楼兰)

The driving force behind the westward movement of the domino effect was not only the newly dominant Xiongnu in the grasslands but also the Central Plains dynasties, which had undergone the trials and tribulations of the Spring and Autumn Period and the Warring States Period and were no longer shallow in knowledge. The anti-expansion to the west began during the Qin Dynasty, even when it was still the state of Qin.

Archaeologists discovered a tomb of laborers from the Qin Dynasty in Lintong, Shaanxi. These laborers were likely involved in the production of terracotta warriors and horses. Through DNA testing of the remains excavated from the tomb, molecular biologists found that most of these laborers had the genetic lineage of the eastern part of the Eurasian continent. However, several laborers had genetic lineages belonging to the western part of the Eurasian continent, with genetic ties tracing back to Central Asia and Iran, among other places. This indicates that the territory of the Qin Dynasty, and even Qin State before it, had already integrated descendants from many Western Eurasian ethnic groups. The exchange of people and genes had occurred quite early.

Of course, this example only illustrates the eastward spread of the Western Eurasian genetic lineage. So, what was the scenario for the westward spread of the Eastern Eurasian genetic lineage?

Foreign scholars once conducted DNA testing on dozens of ancient human remains unearthed in Central Asia. They found that around 2,700 years ago, the people in Central Asia did not possess the genetic lineage of Eastern Eurasia. This suggests that before the Eastern Zhou Period in China, the ethnic groups from the eastern part of Asia had not substantially penetrated into Central Asia. Whether it was the eastern grassland ethnic groups or the forces in the Central Plains, they might not have had the opportunity or capability to move westward.

However, after that period, with the rise of various powerful entities in the Eastern Asian grasslands and the Central Plains, the phenomenon of eastern ethnic groups infiltrating westward began to emerge.

In a site in the central part of Xinjiang, at the foothills of the Tianshan Mountains in Jing County (静县), scholars analyzed the DNA of nine human skeletons. They found that the genetic lineage of seven individuals belonged to the Western or Siberian types of the Eurasian continent, while the genetic lineage of two individuals belonged to the East Asian type. This site dates back to 2,500–3,000 years ago. Additionally, the analysis of human bone DNA from burial sites in the Turpan Basin in Eastern Xinjiang, from the Bronze Age to the Iron Age, also indicated that approximately two-thirds of the tomb owners had genetic lineages from the Western Eurasian continent, while about one-third had genetic lineages from the Eastern Eurasian continent.

These analyses demonstrate that starting around 3,000 years ago, Eastern Asian populations began slowly infiltrating the Western Regions. While the Western Regions still predominantly consisted of Western Eurasian populations genetically, the presence of Eastern populations could not be ignored.

Interestingly, in the Yuansha site (圆沙古城) in the center of the Taklamakan Desert dating from 2,000 to 2,500 years ago, DNA analysis of the tomb owners showed that these people had the closest genetic connection to populations in Southern Central Asia and the Indus River Valley. It appears that while Eastern Asian populations were moving westward, populations from the ancient Indus civilization were also moving northward, with a small number of people entering what is now Southern Xinjiang.

Overall, before the Han Dynasty, the westward movement of Eastern Asian populations was relatively scattered and not as forceful as the westward migrations of western Eurasian populations. Starting with the Han Dynasty, as the Eastern Asian empires expanded, the scale of the westward movement of eastern populations grew larger, and the genetic makeup of various populations in the Western Regions became increasingly diverse.

For instance, Loulan, an ancient kingdom located in the northern part of present-day Ruoqiang County (若羌县), Xinjiang, was the first significant Western Regions power encountered by the Han Dynasty as they ventured westward beyond the Yumenguan Pass (玉门关). In the eyes of the people at that time, leaving the Yumenguan Pass and Yangguan Pass (阳关) meant stepping out of the sphere of Han civilization into a foreign land. From a genetic perspective, when the Loulan Kingdom was first established, the Loulan people may have indeed been quite different from the Han people of the Central Plains. Geographically, the Loulan people lived in the vicinity of the Lop Nur, and as mentioned earlier, the ethnic group of the "Xiaohe Princess" also lived in the Lop Nur area. Some even referred to the people from the Xiaohe Cemetery as the ancient Loulan people. However, the people from the Xiaohe Cemetery lived 4,000–3,500 years ago, and it is difficult to determine whether they had any genetic relationship with the later Loulan Kingdom population.

The name of the Loulan Kingdom first appeared in history in a letter written by Modu Chanyu, the leader of the Xiongnu, to Emperor Wen of the Western Han Dynasty. In the letter, Modu mentioned that he had sent his Righteous King to the west to defeat the Yuezhi, slay their soldiers, and subdue 26 nations, including Loulan and Wusun, bringing them under Xiongnu rule. This unified all the "archery people" into one. From this letter, we can infer that Loulan was originally under the protection of the Yuezhi, but after the Yuezhi were defeated by the Xiongnu, they became subjects of the Xiongnu. This period, during the reign of Emperor Wen of Han, dates back more than 2,100 years ago, making it about 1,500 years earlier than the era of the "Xiaohe Princess."

However, based on historical analysis, before the Xiongnu defeated the Yuezhi, the frequency of eastward migration by people from East Asia to the west should have been low. Therefore, the original population of Loulan might have primarily consisted of the genetic makeup of populations from Western Eurasia. However, as the Xiongnu expelled the Yuezhi and gained control over various Western Regions, it is likely that the population of the Xiongnu confederation would have started moving westward, either through intermarriage, military garrisons, or trade. As previously mentioned, most Xiongnu belonged to the genetic makeup of Eastern Eurasia, so their genetic characteristics would likely have mixed with the population of Loulan.

The Han Dynasty, engaged in a struggle with the Xiongnu, would not have allowed the Xiongnu to control the Western Regions uncontested. Loulan, as the eastern gateway to the Western Regions, became the primary target for the Han Dynasty's efforts to gain influence and control. Starting from the reign of Emperor Wu of Han, the Han Dynasty actively established agricultural colonies and stationed troops in the Western Regions to counter the Xiongnu. Archaeological findings include over 70 Han Chinese wooden documents from a site, which record information about the origins and living conditions of soldiers stationed there. For instance, one document mentions, "里公乘史隆家属畜产衣器物籍" (Li Gong Cheng Shi Long's family members, livestock, clothing, and utensils). "公乘" (Gong Cheng) represents a noble rank in the Han Dynasty, and "史隆"(Shi Long) is a person's name. This implies that officials sent to the Western Regions for agriculture were accompanied by their family members and personal belongings, and they would have certainly worked and lived alongside the local population, leading to mutual integration.

According to historical records, a Han general named Suo Li led over a thousand soldiers from Jiuquan and Dunhuang to establish agricultural colonies in Loulan. They also gathered people and horses from several Western region states and diverted river water for irrigation. As a result, "large fields were cultivated for three years, yielding a harvest of one million measures of grain, and foreign states were subdued (大田三年，积粟百万，威服外国)." The Han Dynasty's mobilization of troops and civilians to establish agricultural colonies in the Western Regions was an important means of control and resistance against the Xiongnu. In this process, the genetic characteristics of people from the Central Plains would inevitably have mixed with the population of Loulan.

In addition to the indirect influx of people from East Asia due to the actions of the Xiongnu and the Han Dynasty, some groups that were originally active in the northwest region would have also gradually intermingled with each other over the course of history, such as the Qiang people. Since they were able to merge with the Yuezhi to form the Southern Mountain Qiang, they would likely have integrated with the residents of Loulan as well. Archaeologists have found a Han Chinese letter from the Wei-Jin Period in Loulan, written by a female Qiang person living in Loulan who was missing her relatives from afar. This serves as direct evidence of Qiang people assimilating into Loulan.

Under the efforts of the Han Dynasty and, to some extent, the Xiongnu, Loulan prospered with abundant resources. By the Eastern Han Dynasty, it had changed its name to Shanshan (鄯善) and expanded in the eastern part of the Western Regions, incorporating places like Xiao Wan, Chadoma, Ronglu, and Qiemo (小宛、精绝、戎卢、且末为鄯善所并). Unfortunately, this marked the peak of Loulan's glory. The fragile natural environment of the Western Regions, unlike the fertile plains of the Central Plains, was not suitable for large-scale agricultural development. The extensive use of water for irrigation by the Han Dynasty gradually degraded the ecological environment of Loulan.

By the time the Eastern Jin Dynasty's eminent monk Faxian (法显) traveled to India in search of Buddhist scriptures, passing through the Lop Nur, it had already become a desolate sea shrouded in yellow sand. In Faxian's "Record of Buddhistic Kingdoms (《佛国记》)," he wrote, "Above, there were no flying birds; below, there were no running beasts. Gazing as far as the eye could see, there was nowhere to be found. I could only use dead men's bones as landmarks (上无飞鸟，下无走兽，遍望极目，欲求度处，则莫知所拟，唯以死人枯骨为标识耳)." The once-thriving Loulan had completely disappeared. Where did the last people of Loulan go?

The centuries-long struggle between the Han Dynasty and the Xiongnu indirectly led to the prosperity of the Silk Road, which traversed the Western Regions, and stimulated the assimilation of various ethnic groups in the Western Regions. Prior to this, trade and ethnic exchanges mostly occurred through the northern grasslands, as transportation was more convenient there. However, the long-standing rivalry between the two Eastern Asian powers obstructed trade and ethnic interactions through the grasslands. Profit-seeking merchants and missionary monks had to opt for the somewhat challenging but still navigable Silk Road, traveling along the two routes north and south of the Tianshan Mountains and the east–west route at the northern foot of the Kunlun Mountains.

In a sense, the blockade of the grassland route spurred the flourishing of the Silk Road and injected a substantial amount of genetic diversity from the east into the Western Regions. Over thousands of years, the continuous eastward migration of Western Eurasian populations was followed by the reverse westward migration of Eastern Eurasian populations, along with limited northward migration by populations from Southern Central Asia and South Asia. Together, these factors contributed to the rich tapestry of human history that earned Xinjiang the reputation as a museum of world races.

6.4 Qiang and Xizang people Hand in Hand on the Plateau

When the Qin and Han Dynasties expanded westward, the first inhabitants they encountered in the northwest region, especially along the Hexi Corridor, were the Qiang people, who grazed horses, cattle, and sheep. Historically, the Qiang people had a wide distribution, primarily centered around Qinghai and Gansu. Before the Zhou and Qin people entered the Central Plains, they had coexisted with the Qiang people in the northwest region.

In the upper reaches of the Yellow River, there is an important tributary called the Huangshui River (湟水). The Huangshui Valley was once considered a geomantic treasure in ancient times. Archaeological excavations in this region have revealed the Kayue Culture (卡约文化), dating back more than 3,000 years to the Bronze Age, which corresponds to the Western Zhou Period in the Central Plains. This culture likely belongs to the ancient Qiang people. According to DNA analysis of human remains found at the site, their maternal genetic lineage is closer to that of modern ethnic minorities in Southwestern China, particularly the modern Pumi people (普米族) in Yunnan, rather than the Han Chinese.

In another cultural site dating from 1,900 to 1,700 years ago in the Huangshui Valley, DNA analysis of human remains suggests a closer genetic relationship to modern Han Chinese, as well as to the Xizangan and Naxi (纳西) ethnic groups in Yunnan. This site corresponds to the late Eastern Han to Three Kingdoms period in the Central Plains. Historical records indicate that during the late Eastern Han period, the Qiang people in the northwest region caused significant trouble for the Han Dynasty, leading to repeated military campaigns against them. This process saw the rise of warlords like Dong Zhuo (董卓) and the beginning of the turbulent Three Kingdoms era, involving figures like Cao Cao (曹操), Sun Quan (孙权), and Yuan Shao (袁绍).

The genetic information from these two sites at different time periods sheds new light on this history. The Qiang people were not a purely homogenous group but rather a collective term used by the Central Plains dynasties to refer to various ethnic groups in the northwest region. During the Western Zhou period, it appears that the Qiang people had not yet been significantly influenced by the Central Plains dynasties. However, as the powerful Qin and Han Dynasties expanded westward, the Qiang people faced the choice of whether to integrate into these dynasties. Some Qiang groups were incorporated into the Qin and Han Dynasties, likely leading to genetic and cultural fusion between the Qiang and Han peoples.

In the late Western Jin Dynasty, Qiang chieftain Yao Chang (姚苌) established the Later Qin regime and competed for control of the Central Plains with various other forces. At that time, the ethnic minorities were not necessarily foreign invaders of the Central Plains; many of them had already been integrated into the system of the Qin and Han Dynasties. There was often genetic intermingling between these ethnic groups, making it more accurate to consider them as cultural rather than genetic groups.

Another group of Qiang people may not have entered the system of the Qin and Han Dynasties for various reasons, such as deep-seated hostilities or cultural differences. These Qiang people embarked on migration routes, potentially heading south to avoid powerful adversaries. They eventually became the ancestral groups of modern ethnic minorities in the southwestern region, including the Yi (彝族), Bai (白族), Hani (哈尼族), Naxi (纳西), Lisu (傈僳族), Lahu (拉祜族), Pumi, and others.

Some Qiang people also migrated westward into the Qinghai-Xizang Plateau, becoming one of the ancestral groups of the Zang people.

The ancestral migration of the Zang people to the Qinghai-Xizang Plateau was facilitated by the cultivation of northern crops. Archaeological findings indicate that farming communities carrying foxtail millet and proso millet settled extensively in the low-altitude river valleys in the northeastern part of the Qinghai-Xizang Plateau approximately 5,200 to 3,600 years ago. Later, the cultivation of naked barley, also known as highland barley or "qingke" (青稞), appeared, providing the ancient inhabitants with a new weapon to occupy the highlands.

By comparing the genetic distribution of two distinct genotypes within modern Xizangan populations and surrounding ethnic groups, molecular biologists have found that the distribution patterns of these genotypes align well with the spread of foxtail millet and proso millet from east to west. Additionally, these patterns have genetic connections to populations found in the Yangshao Culture sites. This suggests that the earliest Xizangan ancestors may have originated from the Yangshao Culture, dating back to at least 4,000 to 5,000 years ago, when they began cultivating foxtail millet and proso millet and eventually migrated to the highlands of the Xizangan Plateau. This migration process involved an expansion from the low-altitude Qinghai River Valleys to the high-altitude Xizangan Plateau.

Around 3,300 years ago, the ancestral Xizangan populations had widespread possession of these two genetic genotypes. During this time or slightly earlier, possibly during the Kayue Culture Period, they started cultivating highland barley and carried it to settle extensively in the snowy highlands.

Among these pioneering settlers were descendants of those who cultivated foxtail millet and proso millet, and the Qiang people also played a role in this process. Furthermore, genes from Central Asian and Siberian ethnic groups also fused into the highland populations. These groups, similar to the Qiang people, may have migrated to the Qinghai-Xizang Plateau in the context of the expansion of the Xiongnu and the Han Dynasty, seeking new living spaces. It's worth noting that barley originally spread to East Asia along with wheat, and Central Asian and Siberian ethnic groups had been cultivating barley for a long time.

Moving further back in time, about 2,000 years ago, the population in the high-altitude regions of the Qinghai-Xizang Plateau experienced another significant expansion. This was an internal consolidation process within the plateau and essentially laid the foundation for the modern Xizangan population's genetic distribution.

During the process of settling on the plateau, the early Xizangan ancestors also domesticated a robust companion—the yak. The history of yak domestication can be traced back to about 7,300 years ago, and genetic research suggests that around 3,600 years ago, the domestic yak population saw a significant increase. Corresponding to the archaeological evidence in the Qinghai-Xizang Plateau, prehistoric populations in the plateau experienced two major population growth phases around 10,000–7,000 years ago and 4,000–3,000 years ago. The former might have been hunter-gatherer groups entering the plateau during the Yangshao Warm Period, while the latter clearly belonged to agricultural communities.

The domestication of yaks and their population expansion closely aligns with human activities in the plateau, indicating that ancient people and yaks collaborated in conquering the plateau. While the earliest physical evidence of domestic yaks was unearthed in the Lhasa Qugong site (曲贡遗址) dating back about 3,750 years, the genetic diversity of domestic yaks is richest in the region around Qinghai Lake in northeastern Qinghai, which suggests that the origin of domestic yaks might have been near Qinghai Lake, also the location of the Huangshui Valley. Based on this information, it can be speculated that perhaps the Qiang people were the first to domesticate yaks, as they were known for their expertise in animal husbandry and nomadic lifestyles.

The domestication of yaks was crucial for the permanent settlement of Xizangan and Qiang ancestors on the plateau, allowing them to move away from their previous seasonal survival strategies and adopt a lifestyle focused on highland animal husbandry and nomadism. This greatly improved their technological and labor conditions for agriculture.

In summary, when tracing the ancestry of Zang people, the foundational genetic source comes from the Yangshao Culture people who cultivated foxtail millet and proso millet. This genetic lineage later merged with Qiang people and, to a lesser extent, with small populations from Central Asia and Northern Asia. By the time of the Xizangan Empire, Zang people already exhibited genetic similarities with modern Zang people. Archaeologists have excavated Xizangan noble tombs in Dulan County (都兰县), Qinghai province, finding that their genetic makeup aligns with Eastern Asian types, dispersed among modern Zang people. Despite a time gap of over a thousand years, there is no significant genetic difference.

As mentioned earlier, Zang people possess unique genes inherited from the ancient Denisovan hominins, enabling them to thrive in the high-altitude, low-oxygen environment. It is challenging to determine which group originally introduced these genes—whether it was the Yangshao Culture people, the Qiang people, or populations from Central Asia and Siberia. Initially, this unique gene likely existed in a small percentage of the population, and individuals with this gene had a survival advantage in the highlands. Over time, the prevalence of this gene in the population increased.

Zang people also passed on these unique genes to the “Sherpas of the Himalayas.” Sherpas (夏尔巴人) are renowned for their roles as porters and guides for Mount Everest climbers.

Regarding the ancestry of the Sherpa people, scholars have had different theories. It is generally believed they might have originated from a branch of the ancient Qiang people. In Xizangan, “Sherpa” means “people from the east,” leading some scholars to speculate that they might be descendants of the Tangut Empire (西夏国)’s royal lineage.

Genetic research has shed light on the Sherpa people’s ancestry. Analysis of the DNA of over 500 Sherpas revealed that they share genetic characteristics with Zang people, both in paternal and maternal lineages. This group is not directly descended from the ancient Qiang people or the royal lineage of the Tangut Empire. Instead, they are a relatively recent branch of the Xizangan population, with the separation of the two groups occurring around 1,500 years ago. This period coincided with intense tribal conflicts on the Qinghai-Xizang Plateau and the emergence of the Xizangan Empire. Perhaps the ancestors of the Sherpa people sought refuge in the Himalayan region to escape the warfare. They also inherited the unique Denisovan hominin-derived genes from Xizangan populations, enabling them to navigate the challenging snowy peaks with ease.

The rise of the Xiongnu Empire in the grasslands and the Qin and Han Dynasties in the Central Plains marked the beginning of a westward genetic expansion across the Eurasian continent. The domino effect that had once tilted eastward had now completely reversed course. The Xiongnu and Qin-Han forces were unstoppable, and various ethnic groups and regimes in the northwest region either surrendered or retreated. This wave of expansion did not stop with the downfall of the Xiongnu and the Han Dynasty; instead, it continued to surge in subsequent history. Several grassland empires like the Turks, Rouran (柔然), Mongols, and the Chinese dynasties such as Sui and Tang actively expanded their spheres of influence to the west, igniting ongoing ethnic blending.

The imperial era was not only a time of biological genetic expansion but also a period of cultural genetic expansion. Before this, from scattered tribes to tribal alliances, and eventually to the earliest ancient civilizations, the overall trend was that social organizational forms became increasingly complex, and the diversity of cultural genes across the vast continent correspondingly increased. However, after empires like the Qin-Han and Xiongnu stepped onto the stage of Eurasia, various cultures in East Asia began to witness a scene where the big fish ate the smaller fish, and the diversity of cultural genes gradually diminished as the cultural landscape of the vast region became more uniform.

This is because biological genetic evolution follows Darwin's principle of "survival of the fittest," while cultural genetic evolution follows a principle similar to Lamarck's concept of "use and disuse."

The prevailing theory of evolution today is based on Darwin's theory, incorporating the concept of modern genetics. It describes biological evolution as follows: genetic mutations occur continuously, and nature selects those beneficial mutations for the survival of organisms, preserving and expanding them, while eliminating mutations that are detrimental to survival. The entire population evolves to become more adapted to the environment.

The evolution of human skin color, for example, aligns with Darwinian evolution. About 2 million years ago, as human body hair decreased, humans evolved dark skin genes during that period to protect against the intense African sunlight. This gene expanded due to its adaptation to the environment. However, when modern humans left Africa for regions like Northern Europe and Asia, where sunlight was less intense, dark skin became a disadvantage, and genes for lighter skin allowed for more vitamin D synthesis, becoming more advantageous in these environments. Hence, lighter skin became more common among modern humans in Europe and Asia. Generally, the process of Darwinian evolution is slow.

In contrast, Lamarck's theory of evolution suggests that traits that are used more become advantageous and are preserved and expanded, while traits that are used less tend to degenerate and be eliminated.[3] This is the "use and disuse" principle. From a biological genetic perspective, Lamarck's theory is largely incorrect because using a trait does not change one's genes, and offspring do not acquire new genes that strengthen a trait. In the field of evolutionary biology, Lamarck's theory is often compared to Darwin's theory and criticized.

However, in the realm of cultural genetics, a culture that is used or favored by more people gains an advantage and quickly erodes other cultures by expanding itself, seizing the resources of other cultures, and even transforming the populations of other cultures into its own believers. This has a hint of Lamarck's "use and disuse" principle.

During the westward expansion of the Qin-Han Empire, the political, economic, and cultural characteristics of Chinese civilization were transmitted to the northwest region. Many cultures in the northwest, including the Rongdi Culture and oasis cultures, suffered heavy blows. Some either disappeared completely, leaving only archaeological sites for future researchers to explore their ancient culture, while others migrated to find new homes, preserving their biological and cultural genes.

Living in the imperial era was unfortunate for smaller cultural groups, but this is history, this is the competition for survival in human society, and this is natural selection. Next, we will see that the ancient cultural groups scattered in the mountainous regions of the southern areas had a similar fate to those in the northwest.

---

[3] Jean-Baptiste Lamarck, 1744–1829.

CHAPTER 7

Whence the Hakka Come: The Majestic Southward Tide of Population

Atop the lofty peak, pristine waters gently sway,
Silver pheasants gracefully traverse, in their own way.
Spiny frogs, with agility, dance upon the liquid's grace,
Myriad terraced fields, in their alluring embrace.
高山顶上水淼淼，
白鹇拖拖过山坳。
石拐咕咕水上跳，
万顷梯田尽妖娆。

Deep within the spring-clad wilderness, a symphony resonates, a mountainous chorus, hummed softly by toiling souls along the terraces. Chongyi (崇义), nestled in the bosom of Jiangxi province, leans against Qiyun Mountain (齐云山), the eminent zenith of southern realms. A staggering kilometer's descent, and layer upon layer of fields unfurl upon these inclines, crowned by a zenith of 62 tiers, validating the adage: "As the mountain's heights, so too the water's heights; the water's heights beget the lofty fields" (山有多高，水有多高；水有多高，田就有多高。).

The Hakka (客家人) folk stand as architects of this resplendent tapestry. Centuries past, their ancestral roots struck firm amidst these emerald hills, taming their agrarian expertise, reshaping treacherous slopes into arable splendor. Long residents of Jiangnan's undulating terrain, they're christened "the hill-folk," their cultural heritage revered as a "living, breathing relic of ancient Han civilization."

Yet, why do they don the mantle of "Hakka"? Pray, whence their origins?

Who, pray tell, reigned as the mountains' ancient custodians? And to what distant horizons do their footsteps wend?

7.1 The "Creolization" of Huaxia

Hakka people are renowned for their ancestral creed: "Better to sell the land of forefathers than to barter our mother tongue, and better to offer ourselves than compromise our essence" (宁卖祖宗田，不卖祖宗言；宁卖身，不卖着). The lodestar that defines the Hakka lies in their distinctive dialect, for they consider themselves inheritors of the Han legacy, bearers of an ancient tongue. This venerable language, interwoven with age-old customs, forms the bedrock cherished by those who dwell amidst the craggy peaks.

Hence, to uncover the genesis of these itinerant souls, one must confront a grand query: Whence came the "Han" (汉)?

How did the Hanfolk emerge as a distinct lineage? What birthed the Chinese tongue we know as Han?

The annals reveal a well-known link between the appellation " Han" and the eponymous dynasty. In the waning days of the Qin Dynasty, scions of long-lost nobility, led by the valiant Xiang Yu (项羽), rebelled against the tyranny that had beset the realm, ultimately toppling the Qin. Dividing the spoils among the feudal lords, Xiang Yu bestowed upon Liu Bang (刘邦) the title of Han Wang, who then led his forces to a haven nestled betwixt the Qinling and Daba Mountains—the realm known as Hanzhong (汉中, the middle of Han).

This name, "Hanzhong," draws its essence from the Han River, a vital tributary coursing through the land, converging with the mighty Yangtze. Within this crucible, Liu Bang forged his stronghold, commencing the epic struggle against Xiang Yu—the "Chu-Han Contention" (楚汉争霸)—a contest that would culminate in his triumph and the birth of the grand Han Dynasty. "Han" derives from the hallowed grounds of Hanzhong, thus entwining the destinies of the dynasty, its people, and its ethnic heritage.

History's pages, mostly inscribed by the conquerors, often extol self and demean the vanquished, disregarding the voices of others or those less fortunate. In truth, whether during the eras of Shang and Zhou or the epochs of Qin and Han, the cartographic reflections seldom align with actuality. The dominion of the central powers seldom held sway over the sprawling territories depicted, for those territories housed not just merchants, Zhou progeny, Qin progeny, or Han progeny.

Consider, for instance, the environs encircling the abode of the Eastern Zhou's sovereign in Luoyang, where an amalgamation of diverse tribes coexisted: Man (夷), Yi (夷), Rong (戎), and Di (狄). Whispers of these encounters persist within the annals: "Within the realms of Zhou, one found to the South the Jingman (荆蛮), Shen (申), Lü (吕), Ying (应), Deng (邓), Chen (陈), Cai (蔡), Sui (随), and Tang (唐). To the north, they encountered Wei (卫), Yan (燕), Zhai (翟), Xianyu

(鲜虞), Lu (潞), Luo (洛), Quan (泉), Xu (徐), and Pu (蒲). Westward, they beheld Yu (虞), Guo (虢), Jin (晋), Wěi (隗), Huo (霍), Yang (杨), Wèi (魏) and Rui (芮). While to the east, there resided Qi (齐), Lu (鲁), Cao (曹), Song (宋), Teng (滕), Xue (薛), Zou (邹), and Ju (莒)—all were kin, offspring, and subjects of King Zhou; the rest souls are hailing from tribes of Man, Yi, Rong, and Di."

If we dub the ancient inhabitants of the Central Plains before the advent of the Han Chinese as the Huaxia ethnic group, then the labels of Man, Yi, Rong, and Di were slyly crafted by the Huaxia, residing in the heartland during the Shang and Zhou Dynasties, to belittle the neighboring clans. So, what role did these peripheral tribes play in the genesis of the Han Chinese ethnicity? Did their genes intertwine with the tapestry of the Han Chinese?

Molecular biologists meticulously handpicked eight archaic populations from 5,000 to 750 years past and delved into the intricacies of their mtDNA. These eight cohorts comprised the ancients of the Central Plains, the Taosi ancients of Shanxi, laborers who toiled at the sepulcher of the First Qin Emperor, as well as the ancients hailing from the northwestern and northeastern realms. Even the Xianbei and Xiongnu populations, discovered within the confines of Mongolia, found their place in this mosaic. Collectively, these ancient clans belonged to the Eastern Eurasian cohort. Subsequently, the genetic profiles of these eight tribes were meticulously cross-referenced with 24 modern Han Chinese groups to unearth the similitudes and disparities in their genetic makeup, thereby unveiling the chronicles of the Han Chinese's genesis.

The savants' foremost revelation materialized in the form of all ancient tribes bearing shared genetic profiles with the modern Han Chinese populace. Furthermore, certain individuals solely exhibited such profiles within the modern Han Chinese fold, dissociating themselves from other contemporary ethnic groups. What does this disclose? It attests that all eight ancient clans bear a blood kinship to the forebears of the modern Han Chinese, with some of their ilk becoming part and parcel of the historical currents that shaped the Han Chinese identity. For instance, the ancient tribes from the northeastern expanse displayed shared genetic profiles solely with the modern Han Chinese, affirming their tangible genetic contribution to the Han Chinese legacy.

The second revelation dawned with the Central Plains ancients manifesting the highest proportion of shared genetic profiles with their modern Han Chinese brethren in the northern climes. Following closely were the northwestern ancients, showcasing a relatively substantial degree of genetic overlap. Alas, the Taosi ancients and their Northeastern counterparts, harking back four millennia, exhibited a lesser extent of shared genetic profiles with the contemporary Han Chinese. The Xianbei and Xiongnu populations, in turn, harbored the slimmest proportion of shared genetic profiles with the modern Han Chinese. A striking pattern emerges: the proximity and recency of ancient tribes correlate with their influence on the formation of the Han Chinese; conversely, the distant and ancient clans wielded a diminished impact.

The undeniable truth of the substantial genetic contribution of ancient Northwestern populations to the Han Chinese illuminates the scenes we've depicted in the preceding chapters. Like a cascade of dominos, the southward and eastward movements of the peoples of Southwestern Siberia ignited a west-to-east migration of tribes, setting the stage for a dramatic intermingling. A subsequent surge, spearheaded by the Xiongnu and the Qin and Han Dynasties, surged westward, enveloping the denizens of the northwestern region, the Di and Qiang tribes, within their encompassing framework. It becomes apparent that a multitude of ancient northwestern clans seamlessly fused with the Huaxia tribe, perpetuating their genetic legacy across successive generations, ultimately shaping the Han Chinese ethnicity.

Even the distant nomadic Xiongnu, who roamed far from the Central Plains, share a semblance of their genetic makeup with the contemporary Han Chinese populace. Throughout history, the Xiongnu and the mighty realms of Qin and Han engaged in a symbiotic dance of economic and cultural exchange, entangled in the relentless whirlwind of warfare, and the spoils of their conquests included not only land but the mingling of their people's genes. It is within these encounters, these collisions, that the genes of the nomads of the grassy plains merged harmoniously with those of the Han Chinese repository in the heartland.

In summation, the genesis of the Han Chinese can be traced back primarily to the Central Plains, approximately three millennia ago. Even then, the ancient denizens of the Central Plains bore a striking resemblance in their genetic structure to their modern-day northern Han compatriots. Ancient tribes from all compass points left their indelible mark, imparting their genetic essence, albeit to varying degrees, upon the northern Han Chinese. Naturally, as time wove its intricate tapestry, disparities in genetic makeup emerged between these ancient tribes and the contemporary Northern Han Chinese population.

Delving deeper, it is evident that the forerunners of the Han Chinese ethnic fabric, the Huaxia tribe, were the result of a harmonious fusion, a convergence of myriad tribes from the Central Plains and its surrounding environs. Certain scholars of antiquity even opine that "Hua" (华) and "Xia" (夏) symbolize distinct ancient tribes. During the uncharted epochs of ice and interglacial ages, the denizens of the northern grasslands ventured south, while their southern counterparts journeyed north, their destinies intertwining in the embrace of the Central Plains. Thus, the Central Plains emerged as the crucible, wherein diverse ethnicities coalesced into an East Asian amalgamation, birthing the Huaxia tribe—an ethereal synthesis born of the ancestral spirits that permeated the ancient realms of the Central Plains.

What is "creolization"?

It is an intriguing phenomenon in the annals of language. With the advent of the Age of Exploration, Spaniards, Portuguese, and their European compatriots ventured forth to colonize the vast expanse of the American continent. After the toil of the first wave of settlers, a new generation was born on American soil, bearing the genetic imprint of their European forebears, and even the African strains that coursed through their veins. They were the native-born denizens of the Americas, these progeny. When these children congregated to engage in play and learning, the varied origins of their parents bestowed upon them distinct tongues. Yet, such is the boundless capacity of the human intellect that these children ingeniously found ways to communicate. They employed the languages passed down by their parents while also borrowing vocabulary from their youthful cohorts. They even ventured to create novel lexicons of their own. And so it came to pass that the descendants of those colonizers forged their own distinctive language, a tongue so elusive that even their own kin found comprehension elusive.

This is the essence of Creole—a linguistic tapestry that characterizes the denizens who dwell within the American lands. In locales such as the fertile soils of Louisiana and beyond, Creole Culture seeped into the very fabric of existence. A cuisine emerged, born from the bounties of the land and sea—rice, tomatoes, okra, and the briny delicacies that abound. Such gustatory delights embraced the piquant touch of fiery peppers, a testament to the unique flavors woven into the Creole tapestry.

As disparate cultures migrated from the four corners of the world to the heartland of the Middle Kingdom, the seeds of Eastern creolization were sown. Languages melded, harmonized, and distilled, until a grand opus emerged—a "feast" aptly christened "Proto-Chinese" (原始汉语). As language is an indelible marker of cultural heritage, those who spoke in the ancient tongue of the forebears were not merely disseminating words; they were fusing together their individual traditions, forging a collective cultural identity. And thus, the Han people were born, the very core of the Han Chinese ethnicity.

The evolution from the Huaxia ethnic group to the Han ethnic group with a common cultural identity has been a lengthy process. The Central Plains region was conducive to agriculture and served as the homeland of the Huaxia ethnic group. However, even within this region, multiple ethnic groups coexisted for an extended period. Agricultural communities occupied the plains and valleys suitable for cultivation, constructing cities with a certain level of defensive capabilities.

During the Shang and Zhou Periods, these agricultural regions were scattered across the vast expanse of land, while mountainous forests far from cities and farmlands served as the homelands of hunting and gathering communities as well as pastoralist populations. For example, around the states of Zheng (郑), Wei (卫), and Lu (鲁), there were many Rong tribes. Zheng was once troubled by the Northern Rong in the Taihang Mountains, and Wei had to relocate its capital multiple times due to the threat of the Rong, eventually settling in present-day Puyang, Henan, but there were still Rong people living outside the city.

For the Central Plains region during the Spring and Autumn Period and earlier, it can humorously be described as "pastoral realms encircling agrarian domains."

The coexistence of multiple ethnic groups in close proximity led to intermarriage and genetic exchange among these groups. "The Commentary of Zuo (《左传》)" records a story from this period. Duke Xian of Jin, during the Spring and Autumn Period, had two sons, Scion Chong'er and Yi Wu, whose fiefdoms were in Pu and Qu, both within the territory of Shanxi. Due to internal power struggles, the two sons planned to leave their fiefdoms and seek refuge with the Bai Di (白狄) people. Chong'er eventually lived among the Bai Di for 12 years.

The Bai Di were a nomadic group, and the area they inhabited was actually quite close to Chong'er's fiefdom. The main reason Chong'er chose to seek refuge among the Bai Di was that his mother, Huji (狐姬), was a Bai Di woman. Chong'er, who later became Duke Wen of Jin, was a prominent figure among the Huaxia ethnic group. He belonged to the Ji (姬) clan of nobility. The first hegemon among the Five Hegemons of the Spring and Autumn Period was Duke Huan of Qi, and Duke Wen was the second. Such an important figure within the Huaxia ethnic group was a product of ethnic fusion, and there were countless instances of intermarriage and fusion among ordinary people at the time.

As the transition from the Eastern Zhou states of the Spring and Autumn Period to the Warring States Period occurred, with small states being absorbed by larger ones, the "barbarian" tribes dwelling in the mountainous forests—referred to as "Man, Yi, Rong, and Di"—also accelerated their integration into the political powers of the Huaxia ethnic group, becoming part of the struggle for supremacy among the contending states. From a genetic perspective, the genes of these tribes all flowed into the genetic pool of the Huaxia ethnic group.

However, during the Warring States Period, the main body of the genetic pool undoubtedly remained the Huaxia ethnic group, as it was determined by the large population supported by thriving agriculture.

The intense competition among the Seven Warring States during the Warring States Period drove these states to adopt "agriculture and warfare" (耕战) as their fundamental national policy. This policy actively encouraged agricultural development and population growth, as population equated to both productivity and military strength. Coincidentally, during this time, ironworking technology had been present in the Chinese lands for some time, and iron agricultural tools began to be used, further promoting the expansion of agricultural areas.

At that time, each of the feudal states had established their extensive agricultural regions. For example, the central region around Xianyang, the capital of the Qin State, the central area of Shandong around Linzi, the capital of the Qi State, and the Central Plains around Daliang (大梁), the capital of the Wei State (魏国), were all renowned for their abundant food production, developed handicrafts, and flourishing trade, as recorded by historians.

Simultaneously, Sima Qian praised the Qin State during the Warring States Period as "a land with rich soil and fertile fields stretching for thousands of miles (膏壤沃野千里)." Other feudal states were no less prosperous. For instance, the population of the Qi State was densely concentrated, leading to the saying, "The crowing of roosters and barking of dogs can be heard in every corner, and there are people throughout Qi's four borders (鸡鸣狗吠相闻，而达乎四境，而齐有其民矣)." The Wei State, on the other hand, "had so many farmhouses and huts that there were hardly any places left for grazing (田舍庐庑之数，曾无所刍牧)."

As mentioned earlier, the population supported per unit area in pastoral and nomadic economies was minuscule compared to agrarian economies. Therefore, with population growth, converting pastoral and nomadic regions into agricultural regions became a rational choice for ancient governments.

As a result, the early landscape of the Central Plains, where pastoral realms encircled agrarian domains, disappeared, and the tribes that had previously engaged in animal husbandry, nomadism, and hunting and gathering also vanished. They assimilated into the vast sea of agrarian populations during the Warring States Period. From a genetic perspective, the gene composition of agrarian populations formed the main body of the genetic reservoir, while the genes of non-agrarian populations were like drops of rain falling into a pond.

This pond is the genetic reservoir of the Han Chinese ethnicity. It is primarily composed of agrarian populations from the Central Plains, supplemented by populations from various other tribes engaged in different livelihoods from all directions. As the vast Qin and Han Dynasties were established, marking the beginning of the "Imperial Era," the concept of Han people quickly gained recognition.

After Qin unified the Six States, its power reverberated across East Asia, and the people of the Qin Dynasty were referred to as "Qin people." However, it's evident that not all the remnants of the conquered Six States necessarily identified as Qin people, as they collectively overthrew the Qin Dynasty. Yet, after the establishment of the Han Dynasty, the people of the Han Dynasty referred to themselves or were referred to as "Han people" and gained widespread recognition. Even after the fall of the Han Dynasty, during the chaotic period of the Three Kingdoms, the Two Jin Dynasties, and the Northern and Southern Dynasties, the concept of Han people, with its "cultural genes," was accepted by various ruling powers, even if they were not of Central Plains ethnicities.

As historians have noted, "The name 'Han' as an ethnic group began with Emperor Liu Bang's proclamation of himself as emperor. In the past, there were various nationalities, all mixed into one, and people adopted the name of the ruling dynasty as the name of our entire ethnicity. From that point onwards, despite changes in dynastic names, the ethnic name remained unchanged (汉族之名，起于刘邦称帝之后。昔时民族国家，混而为一，人因以一朝之号，为我全族之名。自兹以还，虽朝屡改，而族名无改)." Although dynasty names constantly changed, the people identified themselves as Han.

However, during the time of the Han Dynasty, the population distribution was heavily skewed toward the northern regions. In the national census of the second year of the Western Han Dynasty (2 CE), the northern regions accounted for 81% of households, while the vast southern regions only accounted for 19%. This population pattern would be disrupted as populations migrated southwards time and again.

7.2 Hakka the Guest: Men as Guest and Women as Host

When discussing the southward migration of Han Chinese, one of the most representative groups is undoubtedly the Hakka people (客家人). Today, there are tens of millions of people worldwide who identify as Hakka. Since the Hakka people are primarily concentrated in regions such as Guangdong, Jiangxi, and Fujian in China, which are close to the coast, a significant proportion of overseas Chinese immigrants are also of Hakka descent, mostly descendants of Hakka ancestors who ventured southward.

As mentioned at the beginning of this chapter, the Hakka people place great importance on their culture, especially their language. This is likely because their ancestors faced tremendous hardships and struggles with nature and hostile neighboring ethnic groups during their migration from the northern regions to the south. Preserving their cultural traditions helped enhance the cohesion and resilience of this ethnic group. This is the power of cultural genetics.

How close is the relationship between the Hakka people and the Han Chinese ethnic group? What was their migration route like? What stories unfolded between them and other ethnic groups in the south?

The answers to these questions lie hidden within the genetic makeup of the Hakka people.

Molecular biologists collected samples from 148 Hakka men from various locations in Changting (长汀), Fujian province, to analyze the paternal genetic heritage of the Hakka. The results revealed a close genetic connection between the Hakka people in this region and the Han, She (畲), and Dong (侗) ethnic groups. In terms of proportions, over 80% of the paternal genetic heritage was derived from the Han ethnic group, while approximately 13% came from the She ethnic group, and less than 7% came from the Dong ethnic group. This means that the majority of the paternal ancestors of these Hakka people were of Han descent, but a significant portion also had ancestors from the She and Dong ethnic groups, whose genetics were integrated into this Hakka society.

However, the Han Chinese constitute a vast ethnic group, and genetic differences exist among Han populations throughout the country. Where did the ancestors of the Hakka people originate from among the Han Chinese? In this study, molecular biologists classified Han ethnic groups into the Central Plains group, East China group, Hubei-Hunan group, and Southwest group. Most other Han populations in Fujian belong to the Hubei-Hunan group, while the Hakka population in Changting belongs to the Central Plains group. Genetically, they are most closely related to Han people from Hebei and Shanxi.

Therefore, the claim by the Hakka people that they are descendants of Han people from the Central Plains is essentially supported by paternal genetic evidence. However, it should be noted that a small percentage of Hakka people do have paternal ancestors from the She and Dong ethnic groups.

The She ethnic group is a complex ethnic group in the south, and scholars still debate their origins. They are relatively closely related to the Miao (苗) and Yao (瑶) ethnic groups. Some speculate that the She people differentiated from the Miao and Yao groups during the Tang and Song Dynasties and spread from Guangdong. As the Hakka people also share genetic similarities with the Miao and Yao groups, it is likely that the Hakka ancestors absorbed some of the She people's genetics.

Based on certain distinctive genetic mutations, it is speculated that the Hakka population in Changting may have originated from the Central Plains, migrated southward into Hubei and Jiangxi, and then entered Western Fujian. During their time in Hubei, they engaged in genetic exchange with the historical Jingman (荆蛮) ethnic group, incorporating some mutated genes. As for the origin of the small `age of Dong paternal genetic heritage, it might have occurred when these Hakka ancestors passed through Jiangxi and interacted with Dong ancestors.

While the Hakka people's paternal genetic heritage is primarily rooted in Central Plains migration, is their maternal genetic heritage also dominated by Central Plains migrants?

Analysis of mitochondrial DNA reveals an interesting story. Among the Hakka people, there is a mutation in mtDNA, and approximately 20% of Hakka women carry this mutation. In comparison, this mutation is found in 10% of Central Plains Han, 15% of She ethnic group, and as high as 30% to 50% in the Miao and Yao ethnic groups in Hunan and Guangxi.

If the original ancestors of this group of Hakka people all came from the Central Plains region, and the subsequent genetic variations they acquired came from the Miao and Yao ethnic groups (we can simplify the issue by setting the proportion of genetic variations from the Miao and Yao groups at 40%), then how many people from the Miao and Yao ethnic groups would need to be integrated for the Hakka population to increase from a 10% genetic variation rate to a 20% genetic variation rate?

Since mtDNA is inherited through the maternal line, if there were 1,000 women among the female ancestors of the Hakka people who left the Central Plains, and 10% of them had this genetic variation, that would be 100 women. To reach a 20% genetic variation rate, it would require the integration of 500 women from the Miao and Yao ethnic groups. Among these 500 women, 40% or 200 of them would have this genetic variation. In this way, within the integrated population of 1,500 people, 300 would have the genetic variation, which precisely amounts to 20%. Based on this calculation, approximately 33% of the female ancestors of the Hakka people would have come from the Miao and Yao ethnic groups.

In reality, considering that the proportion of this genetic variation among She ethnic women is only 15%, even lower than the 20% mentioned earlier, it suggests significant genetic blending between the She ethnic ancestors and the southern-migrating Hakka ancestors. It can be inferred that within the maternal lineage of the Hakka people's ancestors, females from non-Central Plains Han ethnic groups might constitute around half or even more.

Further genetic studies of broader Hakka populations confirm this inference. Looking at the mtDNA genotypes, the proportion of individuals in the Meizhou (梅州) and Changting Hakka groups with some genotypes from southern regions is around 60%, while in the Heyuan (河源) Hakka group, this proportion exceeds 50%. Compared to other Hakka groups, the Heyuan Hakka group has a closer genetic relationship with Northern Han ethnic groups. What's even more interesting is that different Hakka groups have varying maternal lineage genotypes, implying that although Hakka groups absorbed many females from southern regions, they absorbed different southern ethnic groups' females. This variation might be due to different Hakka groups encountering different southern ethnic groups.

Although the Hakka people number in the tens of millions, they are still a minority compared to the Han population in southern regions today. The patterns observed in Hakka genetic research also apply to the broader Southern Han population. When comparing the genetics of Northern Han and Southern Han populations, molecular biologists have found that, overall, there are differences in both paternal and maternal lineages. Looking at the Y-chromosome genes or paternal lineage, the differentiation between Southern Han males and Northern Han males is not very significant. Southern Han males are mainly descendants of Northern Han males who migrated southward. However, when looking at mtDNA or the maternal lineage, there is a substantial differentiation between Southern Han females and Northern Han females. Among Southern Han females, there are more genes inherited from females of Southern ethnic groups than from Southern Han female ancestors.

In a somewhat imprecise analogy, today's Northern Han population and Southern Han population are like "half-siblings with the same father but different mothers." For most people, their "father" (paternal ancestors) is the same, namely, Northern Han males. However, their "mothers" (maternal ancestors) are different; the Northern Han population is descended from Northern Han female ancestors, while the Southern Han population is largely descended from females of Southern non-Han ethnic groups.

Now, we can roughly outline the story of the southward migration of the Han people. In Chinese history, there were three significant waves of migration from north to south, which occurred during the periods surrounding the fall of the Western Jin (西晋) Dynasty, the An Lushan Rebellion (安史之乱) of the Tang Dynasty, and the invasion of the Song Dynasty by the Liao (辽) and Jin (金) Dynasties. During these turbulent times, one group of Han ethnic groups after another was forced to migrate south to escape turmoil. Due to the hardships and uncertainties of the journey, the migration was predominantly led by males. In ancient society, males were considered strong labor and combat forces. After the Han people migrated south and interacted with Southern ethnic groups, they assimilated many females from these groups into their society. This process might have been achieved through intermarriage between ethnic groups or through wars that resulted in the capture of individuals.

Facing Southern ethnic groups that were less populous and technologically advanced, the migrating Han ethnic groups eventually opened up new territories and became the dominant force in the southern regions, supplanting the original inhabitants. Han males made efforts to preserve many cultural elements from their northern ethnic groups, thus maintaining their cultural traditions, or one might say cultural genes, to some extent.

However, if we emphasize genetic details, then it's true that the males in Southern Han ethnic groups are indeed "guests," while most of the females in these groups are actually the "hosts," the native inhabitants of the southern regions.

7.3 The Xizangan-Yi Corridor: Whom Did They Encounter When Crossing Mountains and Seas?

Ancient Han people were neither the sole inhabitants of northern regions nor the only group to migrate southward. As mentioned in the previous chapter, some of the Di-Qiang (狄羌) ethnic groups migrated to the highlands, becoming the ancestors of today's Zang people, while others moved southward from the northwest region, becoming the ancestors of many ethnic groups in Yunnan today.

Linguistically, the Di-Qiang ethnic groups are important ancestors of the vast Xizango-Burman language group today. The Xizango-Burman language group belongs to the Sino-Xizangan language group, encompassing over 300 languages. Today, speakers of Xizango-Burman languages are distributed widely across East Asia, South Asia, and Southeast Asia, covering countries such as China, Nepal, Bhutan, India, Pakistan, Bangladesh, Thailand, Vietnam, Myanmar, and Laos. Within China, Xizango-Burman language group groups are mainly found in provinces like Qinghai, Xizang, Sichuan, Yunnan, and Hunan. According to modern ethnic classifications, Xizango-Burman language group groups include various ethnicities such as the Zang people (藏族), Achang (阿昌族), Nu (怒族), Dulong (独龙族), Menba (门巴族), Lhoba (珞巴族), Yi (彝族), Bai (白族), Tujia (土家族), Hani (哈尼族), Lisu (傈僳族), Lahu (拉祜族), Naxi (纳西族), Jingpo (景颇族), Jino (基诺族), Qiang (羌族), and Pumi (普米族), with a total population exceeding 20 million.

The significant westward and southward migration of the Di-Qiang ethnic groups began during the Spring and Autumn Period. One of the landmark events was the dominance of Duke Mu of Qin, Qin's sovereign.

Duke Mu of Qin, one of the Five Hegemons of the Spring and Autumn Period, married his daughter to Duke Weng of Jin, another hegemon mentioned earlier. However, during Duke Mu's era, his son-in-law, Duke Wen, governed the state of Jin effectively, and Qin was effectively blocked by Jin west of the Hangu Pass (函谷关) and the Xiaoshan Mountains (崤山), making any eastward expansion extremely challenging. Duke Mu of Qin had no choice but to expand westward, leading to the subjugation of many Rong tribes. Even the Zhou Dynasty's king was astonished by Qin's success in the west and rewarded Duke Mu of Qin with gold drums, acknowledging his significant contributions to expanding the Zhou Dynasty's influence. However, if the Zhou king had known that it was the now-powerful Qin that would later lead to the downfall of the Eastern Zhou Dynasty, how would he have felt?

Qin Duke Mu's expansion westward for thousands of miles was a disaster for the Di-Qiang tribes who originally inhabited the area. These nomadic and pastoralist tribes were highly mobile and, in order to escape Qin's military might, began dispersing and fleeing. Large numbers of Di-Qiang ancestors of the Xizango-Burman language group moved southward along what is known as the Xizangan-Yi Corridor, into Sichuan, Yunnan, and other regions. Geographically, the Xizangan-Yi Corridor "extends northward to the Gansu-Qinghai Plateau and southward to the Yunnan-Guizhou Plateau."

Genetic research has found that the southward migration of Xizango-Burman language group groups resembles the migration pattern of Han people, with the Hakka people as representatives. The Y-chromosome genes of southern Xizango-Burman language group groups are very similar to those of Northern populations, indicating that from a paternal perspective, southern Xizango-Burman language group groups did indeed come from northern regions. However, the mtDNA of Xizango-Burman language group groups in the south differs significantly from both Southern and Northern populations, representing a blend of Southern and Northern populations. From a maternal perspective, the female ancestors of Southern Xizango-Burman language group groups have both Northern and Southern origins.

However, unlike the southward migration of the Han people, the southern migration of the Xizangan-Burman language group resulted in the formation of numerous distinct ethnic groups. This may be because the Di and Qiang ethnic groups were originally quite diverse, with significant linguistic and cultural differences between different tribes. In other words, they had not yet undergone the process of "creolization" to form relatively consistent languages and cultures. Therefore, when different tribes migrated southward separately, their cultural characteristics did not become uniform, ultimately evolving into different ethnic groups. Furthermore, in relative terms, the Han population that migrated southward had a significant advantage in terms of population and technology when encountering local Southern tribes. In contrast, the Xizangan-Burman language group groups that migrated southward did not have as significant an advantage when facing local Southern tribes, and as a result, they were influenced more strongly by the southern tribes, leading to differentiation among different Xizangan-Burman language group groups.

For example, the Yi ethnic group today exhibits significant genetic differences between different regions, and they have many linguistic and cultural branches. Similarly, the Naxi ethnic group, whether in terms of paternal or maternal genetic types, has mixed a significant amount of genetic material from Southern populations, indicating that the ancestors of the Naxi people intermarried and blended with the local Southern tribes during their southward migration.

The proportion of Southern population genes mixed with the Xizangan-Burman language group groups that migrated southward is also related to the timing of their ancestors' entry into the south and the circumstances of neighboring tribes.

From a historical perspective, the ancestors of Xizangan-Burman language group groups in Southern Yunnan, such as the Hani, Lahu, and Jinuo, as well as the Tujia in Hunan, may have entered the southern region earlier, while the ancestors of Xizangan-Burman language group groups in Northwest Yunnan, such as the Lisu, Nu, and Pumi, migrated southward relatively later. Genetic comparisons indeed show that Xizangan-Burman language group groups in Southern Yunnan and Hunan have a higher proportion of genes from Southern populations, while Xizangan-Burman language group groups in Northwest Yunnan have a lower proportion of genes from Southern populations. Genetics and history align quite well in this regard.

What's even more intriguing is that, based on mtDNA, Xizangan-Burman language group groups in Northwest Yunnan are genetically closer to Northern populations, while Xizangan-Burman language group groups in Northern Yunnan and Hunan are genetically closer to Southern populations. This seems to imply that the Di and Qiang ethnic groups may have initially migrated together as a whole group, including both males and females, without significant gender-based migration differences. Female members of the Di and Qiang ethnic groups, being part of highly mobile pastoral and nomadic tribes, could have accompanied the larger group and traveled together with cattle and sheep. In contrast, agricultural groups like the Han people might have exhibited gender-based migration differences during their migrations, with a higher proportion of males in the migration groups.

As a result, the more recently migrating Xizangan-Burman language group groups have a higher proportion of maternal genes from Northern populations. With long-term integration with Southern populations, the Xizangan-Burman language group groups that migrated earlier have a higher proportion of maternal genes from Southern populations. In terms of marriage patterns, females were more likely to marry into societies of other ethnic groups, indicating greater mobility, while males found it difficult to integrate into societies of other ethnic groups, suggesting weaker inter-ethnic mobility. This resulted in the Xizangan-Burman language group groups that migrated southward exhibiting a genetic profile of "male 'guest' and female 'host,'" albeit with a different process compared to the southward migration of the Han people.

Conversely, due to the greater mobility of females between different ethnic groups, there is inevitably a historical presence of Xizangan-Burman language group females marrying into local Southern ethnic groups, thereby injecting Northern population genes into the local gene pool. Across vast mountains and serene lakes, the men and women of Xizangan-Burman language group groups encountered their counterparts and formed unions.

7.4 The Myriad Faces of the Baiyue

The southern region was once the earliest area where modern humans entered China, with our direct ancestors having settled in the rivers, lakes, and lofty mountains of the South at least 30,000 years ago. When the era of agriculture arrived, the ancestors of the southern region were the first to domesticate rice, and the domestication of pigs and dogs also mainly took place in the southern region. When the light of culture first shone on the land of China, the Liangzhu Culture stood tall like a peak in the Jiangsu-Zhejiang region around 4,000 to 5,000 years ago. According to one version of the historians' definition of the "Five Hegemons of the Spring and Autumn Period" (春秋五霸), the fourth and fifth hegemons were not from the Central Plains but rather emerged in the southern region of the Jiangsu-Zhejiang water area as King Fuchai of Wu (吴王夫差) and King Goujian of Yue (越王勾践).

Therefore, when ancient Han people and Di-Qiang ethnic groups began their migration to the south, the southern region was not the wilderness as depicted in Central Plains historical records but had its own long-standing culture and history. For example, in the heartland of southern region, in the middle of Hunan, the Meishan culture (梅山文化), also known as the Chiyou Culture (蚩尤文化), has been passed down. Legend has it that this tribal leader who fought in the ancient battle between Yellow Emperor (黄帝) and Yan Emperor (炎帝) eventually retreated to the Daxiong Mountain (大熊山) in the region. When the pursuers of Yan and Huang arrived there, they found the mountain terrain steep and wild animals abundant, fearing an ambush, they ceased chasing Chiyou. According to ancient texts, the Meishan region was inhabited by the Jiuli (九黎), Sanmiao (三苗), and their descendants. During the Han and Three Kingdoms Periods, there were also active groups known as the Changsha Man (长沙蛮) and Wuling Man (武陵蛮) in Hunan.

For various ethnic groups in different historical periods and regions of the southern region, the most representative collective ethnic name is the Baiyue (百越). The most famous historical figure of the Baiyue is associated with the Yue State, where King Goujian of Yue famously endured hardships and eventually defeated his arch-enemy, the Wu State. He then marched north and formed alliances with the Central Plains states of Qi, Song, Jin (晋), and Lu, becoming the last hegemon of the Spring and Autumn Period.

Looking back at history, Zhejiang and Fujian provinces were once the political centers of the Baiyue ethnic group, a fact supported by genetic research. From a molecular biology perspective, the Baiyue ethnic group has its own unique genetic mutations. Based on the proportion of mutated genes in the population, researchers found that the people of Shanghai and Zhejiang have the highest proportion of Baiyue mutated genes, accounting for about one-fourth. This is likely because Zhejiang was once the stronghold of the Yue State, and many modern Shanghainese people have roots in Zhejiang. Following that, the proportions decrease from Guangdong to Anhui, Hubei, Liaoning, Jiangsu, Jiangxi, Hunan, Henan, and so on, displaying a diffusion pattern along the coastline with Zhejiang at the center.

Based on this, we can speculate that during the southward migration of the Han people, some groups in the southern region with Baiyue genes integrated into Han society and then spread throughout the country, especially along the coastal areas.

However, the Baiyue are not the only major local ethnic group in the southern region; another major local ethnic group competing with them is the Baipu (百濮).

Linguistically, the Baipu ethnic group has close ties to South Asian populations. Geographically, the Baipu ethnic group is more concentrated in the southwestern region of China, closer to South Asia. Long before the Han and Di-Qiang ethnic groups migrated south, the Baiyue and Baipu coexisted in the southern region, and their interaction was likely not always peaceful.

In terms of geographical distribution, the Baiyue seem to be descendants of the modern humans who settled along the southeastern coast during the northward migration (the "Coastal Rampage") and the Baipu appear to be descendants of those who resided in the Southwestern mountains and rivers during the inland migration. From the perspective of the migration routes of modern humans, they did indeed divide into two main routes, coastal and inland. However, we must also relate these two migration routes to the formation of the Baiyue and Baipu ethnic groups, which involves complex interactions and assimilation among different southern ethnic groups, similar to how the Northern Han people formed through "creolization" from various groups from all directions.

Perhaps we can say that the Baiyue ethnic group may have a higher proportion of genes from the coastal migration of modern humans (the "Coastal Rampage"), while the Baipu ethnic group may have a higher proportion of genes from the inland migration of modern humans.

There is an intriguing clue worth contemplating, and that is the presence of some Baiyue genetic elements in certain modern populations of Northeast Asia. For instance, more than 30% of the Buryat people in Russia carry distinctive genes originating from the Baiyue, and even among the Japanese, there exists a small proportion of Baiyue-specific genes. This, of course, doesn't imply that some populations in Northeast Asia and a portion of the Japanese are descendants of the Baiyue ethnic group. Rather, they share common distinctive genes, perhaps inherited from the "Coastal Rampage" modern humans. This group of modern humans reached Japan and Northeast Asia by traveling along the coastline. This clue suggests that the original Baiyue ethnic group might indeed be descendants of these "Coastal Rampage" humans.

Analyzing various aspects like history, language, and culture, the Baiyue seem to have had an upper hand in the power struggle in the southern regions. Genetically speaking, populations originating from the Baipu ethnic group often possess more Baiyue genetic components in their DNA. Conversely, weaker branches have little to no Baiyue genetic influence. For example, some branches of the Baipu ethnic group in Southeast Asia have integrated less than 10% of Baiyue genetic elements. Ethnic groups like the Bulang (布朗族) and Wa (佤族), descendants of the Baipu, have small population sizes and lack Baiyue genetic traces.

Another intriguing clue is the linguistic similarity between the Baiyue ethnic group and some minority languages in Taiwan, suggesting a potential shared ancestral group.

Regarding the earliest inhabitants of Taiwan, there have been various theories. One theory suggests that the ancient Ryukyuan (琉球) people migrated southward around 2,000 years ago, another posits that they came from the islands of Southeast Asia, while the third theory suggests they originated from Southern China.

Considering the remarkable migratory abilities of the "Coastal Rampage" around 30,000 years ago, it's possible that modern humans had already discovered Taiwan during their entry into Chinese territory. However, this hypothesis lacks archaeological evidence. Fortunately, ancient human remains have been unearthed in some ancient sites in Taiwan, allowing for genetic analysis, and the genetic makeup of modern Taiwanese minorities can be used for comparative studies. Based on genetic research, the third theory, which states that the ancient people of Taiwan originated from Southern China, has been supported.

Certain minority groups in Taiwan, such as the Tsou (邹人), Yami (雅美人), and Puyuma (卑南人), possess a specific gene that is widespread on the Chinese mainland, including ancient individuals from the Taosi site dating back thousands of years, coastal Han Chinese communities, inland Miao and Yao groups, and even populations in the northwest. This gene has not been found among Southeast Asian populations. Estimations suggest that this gene emerged around 26,000 years ago in the Yangtze River Basin, likely as a result of mutation after modern humans entered the Chinese mainland. This indicates that the ancestors of some Taiwanese minorities migrated from Southern China to Taiwan rather than directly from Southeast Asia via sea routes.

Among today's Taiwanese minority groups, many share the same genes as ancient individuals unearthed at the Tainan site dating back 1,800 years or more. This gene is more prevalent on the Chinese mainland and less so in Southeast Asia. This further supports the idea that the ancient people of Taiwan, as early as 1,800 years ago, were descendants of mainland ancestors and continued to propagate their genes among today's Taiwanese minority groups.

Around 1,800 years ago, or even earlier, was a time when the Baiyue ethnic groups roamed freely in the southern regions, climbing every mountain and traversing every forest. The migration of ancient people to Taiwan must have occurred multiple times, and among the pioneers who crossed the Taiwan Strait, there were undoubtedly members of the Baiyue ethnic groups.

7.5 Southern Chronicles: Forging a Nation Through Iron and Blood

The epic southward migration of Han and Di-Qiang ethnic groups, spanning nearly a millennium, fundamentally altered the ancient economic and demographic pattern of "the North is strong, the South is weak" in China. By the late Tang Dynasty, when the North was severely affected by the devastating An Lushan Rebellion, the South had already caught up with the North in terms of both economy and population. After a brief period of chaos during the late Tang (南唐) and Five Dynasties (五代), the Song Dynasty emerged, and the South had even started to outpace the North, both economically and in terms of population. For example, according to statistics from the fifth year of the Taiping Xingguo (太平兴国) era of the Northern Song Dynasty (980 CE), the number of households in the South reached 3.874 million, accounting for 60% of the total national households, achieving a complete population reversal against the North. From that point onwards, the North never regained its demographic advantage.

Under the threat of warfare, a large population from the North migrated southward, becoming the primary driving force behind the rapid growth of the Southern population and economy. However, if the South originally possessed the natural conditions and growth potential to support such a large population, why did it have to wait for successive waves of Northern ethnic groups to migrate south before gradually achieving significant population and economic growth?

Scholars argue that the answer lies in technology and agriculture. The Northern agrarian society had accumulated advanced farming techniques and equipment, which were brought to the south by migrating populations, thereby tapping into the Southern region's latent agricultural potential. Additionally, during the Song Dynasty, high-yielding Champa rice was introduced from Southeast Asia, providing farmers with superior crop varieties, further boosting agricultural productivity.

This explanation is reasonable. However, when we reexamine the phenomenon of the South's rise through the lens of genetics, we may find more detailed or slightly different explanations.

Scholars often attribute population growth over thousands of years to agricultural progress, but our genes don't necessarily agree with this view. Looking at the m mtDNA geneotype count of global populations, the largest expansion of maternal lineages in the world's population occurred around 15,000 years ago or slightly later, but certainly earlier than the emergence of agriculture.

So, the narrative of the origin of agriculture should be revised: human societies first experienced a substantial population increase, and under population pressure, our ancestors adopted agriculture, which, as it progressed, further fueled population growth. In other words, the initial occurrence was a "population increase" rather than agriculture.

Why did a population increase occur around 15,000 years ago? As mentioned earlier, this period coincided with the end of the Ice Age, when many tribes may have lost their ability to migrate and were forced to settle in place, relying on their own resources. In the days of hunting and gathering, tribal members often lived far apart, especially when male hunting teams would be away for days or even longer. Because they were compelled to settle during the Ice Age, more men and women gathered together, leading to a rapid population increase.

In theory, as the female population rapidly increased, the male population would have increased rapidly as well. According to the natural laws of human reproduction, the approximate male-to-female ratio at birth is 106:100, meaning that out of 206 newborns, there would be 106 boys and 100 girls. Boys are slightly more numerous at birth because during the fertilization process, sperm carrying the Y chromosome have a slight advantage over those carrying the X chromosome.

However, Y-chromosome gene research has surprised scholars. Tracking the global male population expansion through the paternal line, it is evident that populations worldwide experienced significant growth between 6,500 and 2,000 years ago, with population sizes increasing by 10 to 100 times. The period of rapid male population growth does not align with the period of rapid female population growth.

Looking at the geneotype count, the rapid increase in the female population in Asia occurred around 35,000 to 33,000 years ago. This period coincides with the time when modern humans had already established themselves in south and Southeast Asia and were preparing to move north into East Asia, reflecting a population explosion in Asia during that period. Clearly, at that time, the male population also experienced rapid growth, but it did not result in a significant increase in geneotype count. In terms of population growth rates, the period of highest female population growth occurred around 11,000 to 9,000 years ago, consistent with the arrival of the post-Ice Age warm period and the beginnings of agriculture. The period of highest male population growth, on the other hand, was between 3,300 and 1,500 years ago.

Since the population growth of males and females should ideally be synchronous, why do we observe disparities in genetic representation? We must consider not only birth rates but also mortality rates.

During the hunting and gathering era, the mortality rate among males was quite high. While hunting allowed males to showcase their heroic qualities, it was far from romantic. Prey with low aggressiveness could flee swiftly, and those with high aggressiveness could turn the tables during life-or-death struggles, exacting a toll on even the most skilled male humans. These men didn't return home every day with a magnificent deer or wild boar on their shoulders, while their women waited at the doorstep, ready to prepare a feast. The primary concern for homemakers during the hunting and gathering era was whether their men would return home safely.

Even though the agricultural era began a long time ago, human societies continued to rely heavily on hunting, gathering, and pastoral economies to sustain their populations. Male mortality rates remained significantly higher than those of females. This is why, genetically speaking, female genotypes have increased substantially over the millennia, while male genotypes have lagged behind. Many males did not live long enough to pass on their genetic legacy.

Between 3,300 and 1,500 years ago, a change occurred, with a significant increase in male genotypes. This indicates a rapid growth in the male population, which, of course, also implies a simultaneous rapid increase in the female population. This period coincides with a massive southward migration in China, marking an era of rapid population and economic growth in the southern regions. Apart from advancements in agricultural technology and improved crop varieties, what other factors contributed to the rise of the South?

The substantial increase in male genotypes signifies a significant reduction in male mortality rates. Advances in agricultural technology shifted the focus of males from hunting in the wilderness to working in the fields, decreasing their mortality rates naturally. For Southern China, in addition to the significant development of agriculture and population growth, another crucial factor was the use of iron tools.

In the southern regions, the main challenge in ancient agriculture was not the scarcity of plains or low crop yields but the fact that the warm and rainy climate was conducive to rampant wild vegetation growth. In this lush environment, wild trees, grasses, and flowers would quickly overrun the small plots of land painstakingly cultivated by our ancestors, making it difficult for their crops to grow.

Primitive agriculture is often referred to as slash-and-burn agriculture, where forests are burned to clear land for cultivation. However, this method is only suitable for relatively dry northern regions and is challenging to implement in the humid southern climate. In ancient times, igniting trees was not easy because trees in moist environments contained a lot of water and were resistant to burning. Therefore, ancient people had to first chop down and let the trees dry before setting fire to clear a piece of land.

However, before the invention of sharp bronze and iron axes, chopping down trees with stone axes was incredibly difficult. An experiment was conducted in the Amazon rainforest where indigenous people used stone axes to fell a tree with a diameter of 1.2 meters. It took them 115 hours, equivalent to eight hours a day for two weeks. Clearing an acre of land in this manner would require nearly half a year. And don't forget, in that time, many new shrubs would have grown. Using iron axes, it took them only three hours to fell the same tree. To clear an acre of land, they could complete the task in just over a week. In fact, the Amazonians only transitioned to slash-and-burn agriculture and settled living after the Europeans introduced iron axes to the Americas. Before that, they primarily relied on hunting and gathering for sustenance.

The situation in Southern China was quite similar. Only after the widespread adoption of iron tools could the southern regions usher in a new era of agricultural expansion. Iron tools were not only used for clearing land but also for the production of sharp implements for tilling the soil and more.

The art of iron smelting originated in West Asia, yes, West Asia! The specific location was likely in the vicinity of the Caucasus Mountains. Historically, the Hittites of West Asia were a well-established ironworking group, and they made efforts to guard their advanced technological secrets. After the fall of the Hittite Empire over 3,000 years ago, craftsmen dispersed, and the knowledge of iron smelting quickly spread to various parts of the Eurasian continent, including within ancient China. Iron artifacts dating back more than 3,000 years have been found in Xinjiang and Gansu. However, the introduction of ironworking to the Central Plains region of China probably occurred during the Spring and Autumn Period and was likely facilitated through the Silk Road. By the time of the Qin and Han Dynasties, ingenious Chinese ancestors had developed iron casting techniques that surpassed those of the Western world at the time, maintaining this lead globally until the onset of the industrial age.

In this technological context, Northern Han people and Di-Qiang groups continuously migrated southward, using advanced tools to cultivate Southern farmland. This stimulated population growth and economic prosperity in the south. Whether it was the original inhabitants of the south, such as the Baiyue and Baipu groups, or the newcomers from the north, such as the Han and Di-Qiang people, changes in production methods and technology significantly reduced the mortality rates of their male members.

Some might argue that ironworking technology also spurred advancements in weaponry, leading to an increase in male deaths during battles. This is indeed true, but from a genetic perspective, the impact of deaths in combat appears limited and did not deter the overall trend of declining male mortality rates and the population growth resulting from increased food production. In fact, before the Iron Age, inter-group killings among populations were already occurring, as evidenced by the discovery of "Ötzi the Iceman" in the Alps, a man frozen in ice some 4,000 years ago, likely killed by his own kind.

Accompanying the spread of ironworking were language and culture, as the migration of these groups southward involved the expansion of both biological and cultural genes. There was a synchronous phenomenon between the Y chromosomes of the migrating groups and the spread of their languages, whereas mitochondrial DNA did not exhibit a clear synchronous connection to language dissemination. This reflects the migration pattern emphasized throughout this chapter, where male individuals of different groups tended to maintain their languages, making it difficult for them to assimilate into other groups, whereas females from different groups could more easily intermingle, bridging linguistic and cultural barriers.

So, while we often refer to our spoken language as our "mother tongue," from a genetic perspective, our language is, in fact, our "father tongue," passed down through generations by fathers to their children. The epic of the grand southern migration is interwoven with the resilience of the older generations and the inclusiveness of the mothers.

The Column: The Chronicles of Chinese Canine Migration

In the pages of the third chapter of this tome, we embarked on a journey to understand the origins of our faithful companions, domestic dogs, whose roots trace back to the fertile lands of China's Yangtze River Basin. These early canines served not as mere companions but as sustenance in the distant past. But how did these Chinese domestic dogs venture forth from the confines of the Yangtze River Basin to eventually span the breadth of East Asia and even beyond its borders?

Diligent scholars in China undertook a study of 26 ancient mitochondrial gene sequences of domestic dogs. Astonishingly, they discovered that a significant two-thirds of these specimens belonged to a singular haplogroup, a genetic lineage that stretched uninterrupted from the Yangtze River Basin's embrace to the upper and middle reaches of the Yellow River. Among these canine ancestors, unearthed from the hallowed grounds of the Zhejiang Tianluoshan site (田螺山), dating back some 7,000 years, there lay the progenitors of this lineage. It is here, in this hallowed ground, that one might conjecture to be the cradle of these ancient canines. Delving further into genetic analysis, we find a remarkable event around 7,500 years ago—a pivotal moment when the canine population of the Middle Kingdom experienced an unprecedented surge in numbers. In lockstep, these dogs ventured forth, rapidly populating the northern dominions of the Yellow River Basin and the southern realms along the Yangtze River. This canine expansion, mirroring human endeavors, suggests an intricate relationship with agricultural growth and the corresponding surge in food production.

In the epochs that followed, these canines embarked on an odyssey further southward, making inroads into the southern realms of China. Their journey did not halt there, for they extended their reach into the heartland of Southeast Asia, the expanse of New Guinea, the vastness of the Australian continent, and scattered islands across the Pacific. Their epic voyage even charted the farthest boundaries, gracing the Hawaiian Islands in the northern Pacific and the southern shores of New Zealand in the Southern Pacific. Remarkably, this canine migration route aligns harmoniously with the narrative that unfolds in the subsequent chapter—the expansion of Austronesian-speaking peoples—indicating a shared odyssey where these ancient companions journeyed alongside their human counterparts.

Between 7,000 and 2,000 years ago, the land spanning from the Yangtze River Basin to the Yellow River Basin harbored a tapestry of diversity—a mosaic of at least four haplogroups of domestic dogs coexisting, with the one aforementioned standing as the most prominent. However, the annals of time have borne witness to a transformation among domestic dogs over the past two millennia. A "changing of the guard" has quietly occurred within the bounds of the Chinese mainland.

Delving into the annals of history and tracing the southward migration of the Han Chinese and other ethnic groups, most notably the Hakka community, one may discern the narrative underpinning this changing of the canine guard. Over two millennia ago, in the southern domains, the Baiyue and Baipu peoples nurtured the same predominant breed of domestic dogs, a legacy passed down through the ages. Meanwhile, in the northern territories, the genetic composition of these canines had already begun to evolve. Subsequently, the Han Chinese and the Xizangan-Qiang ethnic groups embarked on their journey southward, their baggage including domestic dogs of distinct genetic heritage, hailing from the northern climes. Their advanced agricultural techniques yielded an abundance of sustenance, nourishing not only their growing human populace but also their cherished canine companions.

In time, the canines from the north eclipsed their southern counterparts, giving rise to a new canine hegemony within the southern domains. Before this changing of the canine guard in the southern realms, the once-dominant canine breeds had already charted territories overseas. Consequently, the genetic lineage of Chinese domestic dogs today exhibits conspicuous disparities when juxtaposed with their counterparts in Southeast Asia, Australia, and the Pacific Islands.

Indeed, dogs have remained steadfast allies of humankind throughout millennia, their journeys intertwined with those of their human companions. Yet, it is worth noting that the migration paths of dogs and humans diverge in subtle ways. When the predominantly male Han Chinese populace journeyed southward, accompanied by their canine companions, gender-based preferences for their dogs may not have been at the forefront of their considerations. Consequently, the gender balance among the southern canine population that journeyed alongside them likely mirrored the natural order. Thus, if the forebears of today's southern populace were marked by a "male-guest-and-female-host" pattern, then the progenitors of the contemporary southern canine population could be succinctly described as a pattern of "male guest and female host," primarily originating from the Northern domains.

CHAPTER 8

Across the Azure Seas: The Genetic Vessel Sailing into the Unknown

Easter Island, suspended in the South Pacific's cerulean expanse, stands solitary amidst its watery domain. The closest inhabited isle lies over 2,000 kilometers away, while the nearest continental coastline stretches more than 3,500 kilometers in distance. An aerial view unveils the island's triangular form, its three vertices fashioned by formidable volcanic outbursts, with scattered vents nestled between these mighty fire-spewers. Easter Island, born of fiery convulsions, owes its name to intrepid Dutch navigators who made landfall on that fateful day in 1722, Western Easter. Yet, the island's Indigenous inhabitants have their own designation—Rapa Nui. Enigmatic and enshrined within this realm are colossal stone figures, captivating all who lay eyes upon them. "Rapa Nui," in the native tongue, carries the evocative meaning of "homeland of the stone statues."

The island's original inhabitants belong to a seafaring lineage, the Polynesians, their ancestry entwined with the maritime tapestry. Some scholars postulate that prior to European discovery, these natives encountered the mingling presence of Native Americans who ventured from the vast Americas. Linguistically, these islanders find their place within the expansive Austronesian family, dwelling amidst a constellation of islands scattered across the vast Pacific, conversing in kindred tongues. Before the onset of European colonial expansion, the Austronesian linguistic realm reigned as the most widespread on earth.

Whence did the Polynesians and their Austronesian kin hail? What distant shores birthed Rapa Nui's Indigenous souls? Could tenuous threads connect them to China's distant realms? These enigmas find their resolution in the realm of genetic inquiry.

8.1 From Paradise Isles to Solitary Ocean Atolls

The Austronesian linguistic kinfolk of today's southern islands boast a staggering populace of 250 million, scattered across an expanse so immense it stretches from the far reaches of Rapa Nui Island in the Southeast Pacific to the distant confines of Madagascar Island in Southeastern Africa. To the south lies the enchanting shores of New Zealand, while to the north they grace the landscapes of Taiwan. Indeed, their domain traverses more than half the circumference of our globe. The captivating allure of these island-dwelling tribes has captivated scholars from every realm, beckoning them to unravel the mysteries of their genesis and the epic odyssey that has shaped their history.

The Lapita Culture (拉皮塔文化), an ancient relic of the Pacific's Neolithic Era, held sway over the isles of Melanesia and Polynesia in the South Pacific. Flourishing some 3,500 years ago, these intrepid forebears crafted exquisite pottery adorned with intricate motifs and fashioned splendid trinkets from delicate seashells. Aboard their intrepid outrigger canoes, they embarked on audacious journeys across the vast expanse of the Pacific Ocean. Today, the islands of the Pacific harbor countless descendants of these intrepid Lapita explorers.

Approximately three millennia in the past, the Lapita people ventured to the remote confines of Vanuatu—a jewel nestled in the western heart of the South Pacific. A prevailing theory of yore posited that these seafarers departed from the sprawling continent of Australia and the lush isle of Papua New Guinea, voyaging eastward to settle upon Vanuatu and other Pacific Archipelagos. Yet, in the hallowed recesses of Vanuatu's archaeological sites, scientists unearthed ancient skeletal remains—a trove fit for genetic scrutiny. A meticulous examination ensued, comparing the DNA samples of over 80 ethnic assemblages, amounting to nearly 800 individuals from Asia and Oceania. Astonishingly, the revelations unveiled a profound genetic concordance between Vanuatu's ancestral inhabitants and their Asian brethren, with a notable affinity to the tribes of Taiwan. Thus, the empirical tides washed away the conjecture of Vanuatu's Australian origins, affirming instead an ancestral genesis in the embrace of Asia's embrace.

As per the genetic perspective, the Lapita's maritime odyssey unfurled thusly: their forebears, it seems likely, embarked from the verdant tapestry of Taiwan, traversing the tumultuous waters of the Bashi Strait that meanders betwixt Taiwan and the Philippine Archipelago. Upon entering this maritime realm, their intrepid souls pushed ever southward, until they graced the mighty shores of New Guinea—a land of prodigious scale. Divided by governance today, its eastern vestiges belong to Papua New Guinea, while its western domains are governed by Indonesia. As the Lapita forefathers alighted upon this Pacific titan, ancient tribes, known as the Papuans, had long since ensconced themselves in its verdant embrace. Notably, a quarter of the South Pacific's populace bears the indelible imprints of Papuan lineage.

Ordinarily, when two tribes converge, the fates conspire to blend their genetic fibers. Curiously, however, the ancient Vanuatu denizens' genetic tapestry reveals scant traces of Papuan ancestry. In essence, these resolute Lapita wayfarers, hailing from the isle of Taiwan, forged no emotional bonds with the Indigenous denizens of New Guinea. Rather, they swept past the island's northern fringes with swift abandon, forging eastward toward the Solomon Islands before pressing onward to the remote realms of the boundless Pacific Ocean's isles.

Scholars categorize the migration of Austronesian-speaking tribes, epitomized by the Lapita ancestors, as the "express mode." They resembled maritime nomads, traversing the vast oceanic expanse. Settling upon a newfound island, they eked out livelihoods for generations until resources waned, prompting them to embark on vessels, seeking the next island—akin to nomadic tribes meandering from one grassy plain to another. Unlike the Papuan people, they did not embrace the agrarian life, forsaking settled agriculture. Thus, the chasm in lifestyle and culture hindered any romantic tales or genetic mingling.

Curiously, a more intricate answer to the origins of the Lapita ancestors emerges from a genetic study—a study of pigs, not humans.

Archaeologists unveiled the Lapita's maritime expertise alongside their adeptness in swine husbandry. Superficially, one might surmise that bountiful fish in the ocean would satiate their carnivorous desires, assuming they possessed piscatorial acumen. But reality veers from this notion. Similar to numerous mainland locales bereft of game for hunters, many marine realms lack piscine life or are desolate in fishy resources. Fish migration adheres to established routes and specific habitats; beyond these domains, the ocean mirrors the arid wastelands of the mainland. Within such precincts, the Lapita's porcine rearing skills assumed paramount importance, replenishing their sustenance, notably the precious succulence of meat.

Compounding matters, pigs, unskilled in the aquatic arts, traversed the Pacific in unison with their human counterparts. Scientists procured genetic samples from modern wild boars across varied Pacific regions, discovering that the swine dwelling upon numerous isles were descendants of escaped domesticated brethren. Additionally, genetic material from mainland Asian wild boars and ancient boar tusks gleaned from museum collections were included. Astonishingly, these endeavors unveiled a mesmerizing revelation—all these boars spanning divergent lands bore kinship with a specific Vietnamese wild boar lineage, implying that Vietnamese swine served as the common progenitors for the Pacific's entire porcine cohort.

Does the wild boar's genetic evidence prove that the primordial sailing grounds of the Lapita forebears were not in Taiwan of China, but rather in Vietnam?

The millennia-long maritime odyssey of Pacific islanders may possess an intricacy beyond measure. Some 4,000 to 5,000 years ago, propelled by advancements in nautical prowess, numerous seafaring cohorts likely embarked on ventures into the heart of the Pacific, seeking hospitable isles to call home. Within this diverse assemblage, fortune smiled upon a select few, guiding them to suitable havens where their genetic legacy took root and flourished amidst modern island-dwellers. Alas, others faced a cruel twist of fate, losing themselves irrevocably in the mercurial oceanic clime, with their genes forever silenced, absent from the annals of posterity.

The porcine plight echoes that of humanity, yet there exist instances when islanders vanish or abandon their shores, wherein fate conspires to spare certain swine, allowing them to survive in the wilderness or fall into the hands of more accomplished mariners, reintroducing their genetic heritage to the expansive fold of the Pacific's "isle swine" fraternity.

Further genealogical revelations emerge from porcine scrutiny, unveiling a dispersal of these creatures across two distinct domains: one centered around New Guinea, wherein the swine were potentially conveyed by agrarian predecessors of the Austronesian lineage, who traversed the tapestry of Southeast Asia. The other realm sprawls across Taiwan, the Philippines, and the western fringes of Micronesia, where the swine bear correlation to the migratory exploits of the Lapita folk and their Polynesian descendants. As the Lapita ancestors navigated the pathways encircling New Guinea, they opted not to abscond with the islanders' genetic fabric but instead absconded with their porcine compatriots.

How may this enigma be deciphered? The agrarian Austronesian populace in New Guinea, replete with bountiful swine herds, seemingly eclipsed their Lapita counterparts, who, in their ceaseless maritime perambulations, traded anchorage for adventure. It is plausible that the Lapita yearned for the Austronesians' prized pigs, either through barter or the opportunistic acquisition of swine during unguarded moments, hastily seizing them and fleeing upon their seafaring vessels.

Ergo, the Lapita forebears set sail from the shores of Taiwan, China, while their porcine companions embarked from the reaches of Southeast Asia, specifically Vietnam. On some hallowed isle within the Pacific expanse, man and beast converged, forging an enduring bond amidst their shared journey. This tale, adorned with veracity, unfolds.

Yet another testament to the assertion that the genesis of the South Island language stock traces back to Taiwan, China, emanates from linguistic inquiry. Carved by geographic impediments that cleave islands from one another, the vast linguistic empire of the South Island harbors an astonishing tapestry of up to 1,200 diverse branches. Linguists, striving to unearth patterns amidst this profusion, meticulously organize them into ten linguistic families, nine of which subsist among Taiwan's mosaic of Indigenous tribes. Analogous to Africa's fecund cradle, teeming with genetic variance and renowned as the birthplace of humankind, places abounding in linguistic diversity typically serve as the birthing ground for language families. Thus, scholars, guided by this logic, posit Taiwan as a plausible fountainhead for the South Island language stock, if not a vital crucible of its formation.

Bearing witness to this evidence, the Polynesian men and women, gracefully adorned in grass skirts as they execute their rhythmic dances upon Pacific isles, emerge as kindred spirits with their Chinese brethren. We are kin, bound by ancestral ties that traverse time and space.

8.2 From Nanyang (南洋) We Journeyed, to Nanyang We Strayed

The Peranakan Chinese ancestors' voyage speaks of an ancient intertwining between China and the wider world. Those brave pioneers who set forth from China (including the Taiwan island) merely represented a single thread among the vibrant tapestry of global ethnicities emerging from China's embrace.

When evoking Southeast Asia, the Chinese often conjure the phrase "venturing to Nanyang (下南洋)." "Nanyang" defined the realm of Southeast Asia during the Ming and Qing eras, particularly encompassing the Philippine Islands, the Malay Archipelago, the Indonesian isles, and the coastal enclaves of the Indochinese Peninsula. As China's population swelled during those dynasties, the southern provinces, teeming with souls, compelled a multitude to traverse the seas, seeking livelihoods and fortunes in Nanyang's embrace.

In tracing the migratory paths of our ancestors, we find their first steps etched upon the shores of Southeast Asia, where they paused briefly before progressing northward into East Asia, ultimately finding their way to China. These audacious Coastal Rampage constitute the majority of our forebears, for we, their descendants, owe our origins to the pioneering spirits that embarked upon Nanyang's shores.

From Nanyang we journeyed, and to Nanyang we strayed. The duration of our homeward odyssey exceeds by far the era of Ming and Qing sojourns to the southern lands.

Taiwan, a mere waystation on the Austronesian-speaking ethnic groups' migration, reveals an ancestral chronicle intertwined with the East Asian mainland.

Molecular biologists, through their discerning lens, have unearthed a linguistic affinity between the Austronesian-speaking groups and the Dong (侗) and Dai (傣)-speaking clans of Southern China, accompanied by a shared legacy of distinctive genetic mutations within the Y chromosome. This remarkable genetic lineage can be traced back to the dawn of China's Neolithic Era, where ancient denizens of Zhejiang bore witness to its emergence. Moreover, the recent discovery of an ancient skeletal relic, dating back some 8,400 years in Fujian, provides molecular biologists with further clues. From its genetic essence, they deduce an intimate kinship with the denizens of central and Southern East Asia, most notably the Yami (阿美人) and Atayal (泰雅人) peoples of Taiwan, while their ties to the various ethnic groups of Southeast Asia remain more distant.

Through meticulous genetic analyses and scrutiny of mutation rates, we discern that the ancient Austronesian-speaking clans, forefathers of these wanderers, once thrived within the tapestry of Jiangsu and Zhejiang in China. Then, approximately 5,900 years ago, they embarked upon a momentous migration, an audacious exodus that carried them in great numbers toward the sun-drenched southern realms, both land and sea.

These early forebears birthed diverse clans, branching off into distinct factions. Take, for instance, the gradual evolution of certain groups within the southern reaches of China, molding them into the Dong-Dai linguistic lineage. A faction of this lineage embarked eastward, spanning the sea to settle in Taiwan. After a period of metamorphosis, they pressed further south, establishing the enigmatic Lapita Culture on the Philippine Archipelago and the eastern fringes of Indonesia. From there, their influence rippled across the vast expanse of the Pacific. Simultaneously, another sect of the Dong-Dai linguistic lineage ventured southward, meandering along the mainland until reaching the Indochinese Peninsula. Their journey continued, propelling them westward into the Western isles of Indonesia, ultimately forming the Malay people. Roughly 2,300 years ago, the Malays seeped eastward, infiltrating the shores of New Guinea, where they mingled genetically with the ancient Papuan populace, spreading their cultural sway deeper into the Pacific's abyss.

Picture the entire South China Sea as the Mediterranean, nestled betwixt Europe and Africa. The southern emissaries of the Dong-Dai or, rather, the Austronesian language group, diffused from the north to the south along both sides of the South China Sea, effectively transforming the region and its environs into a sphere of "Mediterranean civilization" in Indo-Pacific region.

Genetic revelations unveil an ancestral exodus, transpiring approximately five to six millennia ago, where ancient inhabitants from China's Jiangsu and Zhejiang provinces exhibited a proclivity for southward migrations toward

Southeast Asia. What catalyzed such a migratory wave at that juncture, in those precise locales?

During that epoch, the northerly Han populace had yet to undertake their sojourn toward the south, thereby freeing the ancestors of the Austronesian language group from the yoke of external oppression. Plausible supposition suggests that this epoch coincided with the dawn of agricultural prosperity. Both millet, flourishing in the north, and rice, thriving in the south, had undergone domestication, augmenting the sustenance available to the agrarian masses across the vast Chinese expanse. Consequently, regions conducive to cultivation experienced an exponential surge in population growth. Notably, the Jiangsu and Zhejiang provinces emerged as seminal sites for the early domestication of rice and flourished as bastions of ancient culture in the southern domain. Thus, this momentous emigration, engendering the birth of the Austronesian language group, likely transpired as a consequence of burgeoning populations.

This conjecture finds validation within the annals of the water buffalo. As alluded to earlier, the domestication of bovine creatures germinated in West Asia and India, with yellow cattle extensively populating the Yellow River Basin roughly 4,500 years ago. Meanwhile, the water buffalo, indispensable to Southern China, played a pivotal role in propelling the cultivation of paddy fields.

Water buffalo may be categorized into two distinct strains: the swamp variant and the river variant. The former, prevalent in Southern China and Southeast Asia, harmonizes with the rice-growing milieu, while the latter finds solace in the lands of South Asia and the Mediterranean region. Thus, the common folk of China primarily encounter the swamp buffalo in their daily lives.

The time of divergence between swamp-type water buffaloes and river-type water buffaloes dates back hundreds of thousands of years, predating the exodus of modern humans from Africa. The precise domestication era of swamp-type water buffaloes, as indicated by genetic data, falls within the range of 3,000 to 7,000 years ago. This initial taming likely occurred in the lower regions of China's Yangtze River. Subsequently, domesticated water buffaloes embarked on two distinct routes: one against the current, tracing the Yangtze River upstream, and the other venturing southward. During their migratory path, several domestication events took place around 4,100 years ago in the upper reaches of the Yangtze River, followed by another occurrence roughly 3,300 years ago, spanning from Southwestern China to northern parts of Southeast Asia. Notably, it was around 3,300 years ago that swamp-type water buffaloes experienced a substantial expansion.

Considering the geographical origins, timeframe, and dispersal patterns of swamp-type water buffaloes, a striking parallel emerges with the domestication and spread of rice across Southern China. Both phenomena arose from the lower regions of the Yangtze River before making their way southward. Hence, as the wanderers of the Austronesian language group embarked on their migration, they carried with them rice seeds and guided their water buffaloes from China's Jiangnan region toward the southern expanses of the country, eventually reaching the Nanyang region.

Of course, the southern trajectory of the Austronesian language group was no simple colonization of virgin lands; it was a profound "journey of roots" for wanderers who had long been adrift. These were the ancient coastal nomads who traversed Southeast Asia tens of thousands of years ago. As the Austronesian language group returned to Southeast Asia, they encountered kindred spirits—descendants of the coastal nomads who had resided in the region for generations. Thus, these seekers of origins, arriving adorned with precious gifts of domesticated rice and accompanied by stalwart aides—the water buffaloes—undoubtedly shared in the delight of introducing flavorful companions, such as pigs, dogs, and quite possibly the chicken, to the lands of Southeast Asia.

Now, let us turn our attention to the enigmatic origins of the chicken, if only briefly. The turkey devoured by Westerners on Christmas traces its lineage to the American continent, a matter we need not explore here. The familiar domestic chicken, however, descends from the untamed red junglefowl, and scholars continue to engage in spirited debate over its birthplace. Archaeological revelations beckon us to the Yin Ruins, where chicken bones unearthed from 3,300 years ago have undergone meticulous scrutiny, thus affirming their domestic heritage. The oracle bone inscriptions at Yin Ruins distinctly differentiate between the characters for "chicken" (鸡) and "pheasant" (雉), with the former signifying domestic fowl cherished for sustenance and ceremonial rites, while the latter denotes the capture of wild fowl.

As for the earliest substantiated remnants of domestic chickens within China's borders, they emerged from the earth in Inner Mongolia, a testament to their presence 3,600 years ago. Among the scholarly pursuits in unraveling the past, mtDNA analyses have illuminated a shared genotype among northern Chinese chickens, its fullest diversity manifesting in the Central Plains—an origin that likely took root and spread across the northern territories some 2,500 years ago. Curiously, the annals of the Commentary of Zuo (《左传》), recounting tales of ancient China, document a momentous clash of roosters between the Jishi (季氏) and Youshi (郈氏) families, an event that synchronizes intriguingly with the northward expansion of domestic chickens.

However, in contrast to the domestication of numerous livestock and poultry, these time periods clearly prove to be too tardy, rendering them highly implausible as the earliest moments of chicken domestication. The red junglefowl, indigenous to the realms of Southeast Asia, primarily inhabits the sultry tropical regions south of the Tropic of Cancer. Hence, its initial domestication, akin to that of swine, plausibly transpired within the Indian subcontinent's fertile Indus River Basin, subsequently permeating the landscapes of Southeast Asia. Accompanied by *Homo sapiens* of modern intellect, this journey led them across the continent, wherein they underwent repeated taming amidst the blossoming agrarian society until the specific breeds ultimately materialized. Consequently, as the Austronesian language-speaking tribes flourished, the gallinaceous poultry found themselves sharing in the good fortune.

From a geographical standpoint, the Austronesian linguistic cohorts seemingly displayed a penchant for nautical exploration, while their inclination to take root in inland territories was relatively lackluster. Within the continental expanse of Southeast Asia, their presence remained confined, merging predominantly with the Cham ethnic populace as evidenced by linguistic and genetic assimilation. The Cham people, predominantly situated within the central and southern reaches of Vietnam, once played a pivotal role in the establishment of the Champa Kingdom, an ancient bastion of civilization within the Southeast Asian domain.

An examination of their matrilineal mtDNA unveils an amalgamation between the Austronesian-speaking tribes and the Indigenous Mon-Khmer populace. Meanwhile, a scrutiny of their Y chromosomes reveals an intricate amalgam derived from the gene pools of the Malay Archipelago, the Indochinese Peninsula, and even South Asia. Historical narratives trace the roots of the Champa Kingdom to the tenets of Hinduism, fostering a close alliance with South Asia, wherein select religious and commercial emissaries hailed from the southern realms.

Alas, the Austronesian-speaking tribes failed to venture deep into the heartlands of Southeast Asia, their expansionary strides invariably thwarted by the formidable might of the Indigenous Mon-Khmer people.

8.3 Ancestral Homeland of the Japanese: Across the Seas

"In the depths of the sea, familiarity abounds, bridging distant shores" (海内存知己，天涯若比邻), as the Austronesian tribes transformed the South China Sea into their private haven, a surge of transoceanic expansion took hold among the ancient inhabitants of Northern China.

When it comes to our neighboring nation, Japan, the Chinese often evoke the phrase "a mere strip of water" (一衣带水), depicting the narrow expanse that separates our lands. Geographically speaking, the Yellow Sea and East China Sea stretch wide between our coastlines, dispelling any notion of a meager divide, which historically burdened the flow of our people. Yet, "a mere strip of water" resonates on a deeper level, evoking the cultural and intellectual affinity between our two nations.

In contemplating the ancient intermingling of our peoples and the exchange of their genetic heritage, the metaphor of "a mere strip of water" befits their journey. Confronted by vast seas, the ancient mariners were confined by limited nautical prowess, yet they compensated by traversing time rather than distance. With eons at their disposal, they ventured forth, gradually transforming the expanse of the Yellow Sea and East China Sea into a narrow conduit.

The ethnic tapestry of the ancient Japanese Archipelago bears an indelible mark of its origins within China's sprawling territories. Primarily, the eastern and southern realms of the Japanese islands gaze upon the boundless Pacific, while the closer proximities of the Korean Peninsula to the northwest and the Shandong Peninsula to the east beckon. The northern reaches of the Korean Peninsula precisely align with the northeastern regions of present-day China. Moreover, the *Homo sapiens*' exodus from Africa unfolded west to east, south to north, necessitating passage through the vast expanse of China before embarking upon the seas toward the Japanese isles.

When contemplating the lineage of the Japanese forebears, the Chinese mind instinctively conjures the legend of "Xu Fu's Eastern Odyssey" (徐福东渡). Emperor Qin Shi Huang dispatched Xu Fu on a quest for the elusive elixir of eternal life, accompanied by young lads and lasses who embarked upon the fathomless sea. Coincidentally, during the tumultuous era between the third and second centuries BC, a profound transformation swept through the ethnic makeup of Japan. Thus, the threads of destiny intertwined these two events, weaving a narrative of Xu Fu's eastern sojourn imparting not only Chinese culture but also the very essence of its people, irrevocably reshaping the primordial state of Japan at that time.

However, in terms of the dispersion of modern *Homo sapiens*, the ethnic bond between China and Japan surpasses the ancient era of Emperor Qin Shi Huang. Countless eons ago, the intrepid "Coastal Rampage" hailing from Southeast Asia cascaded upon the southeastern shores of China, propelled by an unrelenting impulse to expand northward. Unyielding to the elements, they pressed onward, traversing a mere 40-kilometer-wide channel called the Tsushima Strait, nestled betwixt the Korean Peninsula and Kyushu Island in present-day Japan. In the depths of the glacial epoch, when the sea levels had plummeted, baring vast stretches of the seafloor, the strait constricted, offering no impedance to the ancient seafarers of yore. Thus, it is highly plausible that the valiant Coastal Rampage reached the realms of the Korean Peninsula and the Japanese

Archipelago, thus becoming the vanguard of the bygone era, migrating from China to Japan.

Ancient relics, such as unearthed stone tools, attest to the existence of human habitation in the Japanese Archipelago some 30,000 years ago. These stone implements predominantly served the purpose of hunting. During this epoch, a time when the Shandingdong Man roamed the Dragon Bone Hill vicinity in Zhoukoudian, one can muse that these early Japanese ancients might have descended from the "steppe hunters" that traversed the icy lands. Perchance, their genetic lineage bears proximity to that of the Shandingdong Man. Regrettably, the scientific community has yet to unearth the remains of these prehistoric individuals, leaving genetic analysis as an unfulfilled aspiration.

The earliest vestiges of ancient humans within Japan's borders were discovered amidst the splendor of the Ryukyu Islands, bearing testimony to an existence dating back some 18,000 years. Alas, a lamentable scarcity of genetic analysis plagues this population, leaving us solely with the medium of skeletal morphology for comparative analysis. Given the Ryukyu Islands' tropical embrace, ensconced within the vast sea, distanced by no fewer than 300 kilometers from the mainland's Honshu Island, it stands to reason that, during the glacial age, no land bridge tethered the Ryukyu Islands and Kyushu Island. Geographical barriers proved formidable, rendering these primeval Ryukyuans as progeny of the intrepid Coastal Rampage, journeying from Southeast Asia, hugging the southeastern shores of China.

Approximately 16,000 years prior, during the twilight of the Last Glacial Period, the Japanese Archipelago witnessed the arrival of a novel cohort of ancient souls. Earning their keep through the art of hunting, fishing, and gathering, they bore the remarkable ability to craft pottery adorned with intricate rope-like patterns. Thus, bestowed upon them was the moniker "Jōmon people" (绳纹人), while their cultural ethos was dubbed the "Jōmon Culture."

From whence did these Jōmon people emerge?

The Jōmon Culture, stretching back some 16,000 years and fading around 2,300 years ago, boasts a staggering timeline surpassing millennia. It marks the transition from the ancient epoch of hunting and gathering to the dawn of agriculture. One might assume that over this expanse of time, the Japanese Archipelago would have experienced numerous interminglings with neighboring continental populations. Yet, upon scrutinizing the Jōmon people's way of life, it becomes apparent that they never gained access to the art of farming or the cultivation of suitable crops. While they seemingly made forays into taming wild flora, their efforts yielded little substantive progress.

One can surmise that not long after the Jōmon folks migrated to the Japanese isles, the Last Glacial Period relinquished its grip, as spring resurrected the land and the essence of life. Consequently, the land bridge linking the Japanese Archipelago to the Korean Peninsula succumbed to oblivion. The balm of a warmer climate bestowed vitality upon the Japanese islands, rendering them a veritable cradle for existence. However, the Jōmon people, stranded on these isolated islands, found themselves ensnared in a tragic technological quagmire devoid of foreign genes or the infusion of external expertise.

Might their fortunes plummet further into the abyss?

Let us delve into the Jōmon people's genetic heritage. Japanese scientists chanced upon the remains of a Jōmon-era woman in Hokkaido. Radiocarbon dating suggests her demise transpired between 3,960 and 3,550 years ago, plausibly aligning her existence with that of humans 3,800 years in the past. By extracting DNA from her molars for scrutiny, scientists unearthed a peculiar gene predisposing her to digest high-fat sustenance with ease. Astonishingly, 70% of the Arctic denizens of today bear the same genetic marker. This does not intimate that this Hokkaido dweller belonged to the Arctic lineage, but rather she shared a common ancestry with the contemporary Arctic populace. From a genetic standpoint, this Jōmon woman from antiquity appears to hail from the northern reaches of Asia, sharing blood ties with present-day Indigenous communities residing in the eastern expanses of Russia.

The genetic imprints of this woman also underscore that the progenitors of the Jōmon populace, exemplified by her existence, diverged from other Asian populations some 38,000 to 18,000 years ago, charting an independent evolutionary path. Hence, it stands to reason that during this era of divergence, the forebears of these Jōmon people embarked on a migration across the Korean Peninsula, ultimately finding solace upon the shores of the Japanese Archipelago.

Further Jōmon-era remains were unearthed in Fukushima, Japan. Analyzing the DNA extracted from the dental remnants, scientists identified pronounced genetic disparities between this Jōmon population and their counterparts in East Asia and Southeast Asia. Among the modern Japanese populace, the Ainu people (阿伊努人) emerge as the closest kin to the Jōmon people, followed by the Ryukyuan people (琉球人), and then the denizens residing in the environs of Tokyo.

This revelation may suggest that the Jōmon populace underwent a distinct evolutionary path within the confined boundaries of Japan's Archipelago, devoid of substantial gene flow with mainland communities. Thus, over time, they accumulated substantial genetic disparities.

The scrutiny of modern Japanese Y-chromosomal genes lends credence to this conjecture. Scholars meticulously extracted Y-chromosomal genes from over 300 Japanese individuals, meticulously comparing them to their East Asian counterparts. Remarkably, every Japanese individual possessed a profoundly unique set of genes, tracing back to their Jōmon ancestors.

During Japan's late Jōmon era, a parallel to China's historic Warring States Period, a frigid climate gripped the East Asian region. Consequently, the Jōmon population experienced a drastic decline. Scholars, employing calculations of genetic mutation rates among individuals bearing the distinct Jōmon Y-chromosomal genes, extrapolated ancient population figures. Astonishingly, from approximately 3,200 to 2,000 years ago, the Jōmon population plummeted from its zenith of 260,000 to a mere 80,000. Although the Jōmon community fared better in comparison to the island's Indigenous inhabitants, numbering only in the thousands, an unpromising future awaits them unless circumstances change, leaving them precariously poised on the precipice of a technological tragedy akin to Easter Island.

Climate change, for the Jōmon people, possesses a double-edged nature—both a blessing and a curse. In their arduous struggle, a sudden influx of resolute "new" individuals traversed the land bridge.

The annals of antiquity unveil a significant transformation in ancient Japan during the third century BCE—a watershed moment marked by a novel ceramic style, the sudden advent of bronze and iron implements, and the cultivation of rice. Locating the epicenter of this "new" epoch in a place called Yayoi-cho, Tokyo, scholars christened this migrating collective as the "Yayoi people" (弥生人), encapsulating their cultural tapestry as the Yayoi Culture.

Even in terms of physicality and dental characteristics, a palpable distinction emerged between the incoming Yayoi contingent and the venerable Jōmon ancestry. The Yayoi folk boasted towering physiques, their cranial profiles diverging from their Jōmon predecessors. Notably, dental morphology reflected their dissimilarity—while the Jōmon populace predominantly sported the telltale shoveling of incisors, the Yayoi people exhibited sinodont teeth, reminiscent of their Chinese roots. As elucidated in Chapter 2, the advent of shovel-shaped incisors predates their sinodont counterparts, diffusing along East Asia's coastal fringes. Conversely, sinodont teeth emerged later, disseminating across East Asia and the northern realms of the Asian mainland.

Genetic analysis reveals that the Yamato people (大和人), the predominant ethnic group in modern Japan, carry approximately 12% of their genetic heritage from the ancient Jōmon folk, while the bulk of their DNA can be traced back to the ancient Yayoi population. When juxtaposed against continental communities, the Yamato people display closer genetic affinities with the Han Chinese residing in the northern territories and select ethnic minorities inhabiting the northern fringes. Such genetic investigations not only underscore the Yayoi people's arrival on the Japanese Archipelago from the northeastern realms of China but also substantiate their passage through the Korean Peninsula.

The epoch when the Yayoi people settled in the Japanese isles coincided with a tumultuous chapter in ancient China's annals, the period stretching from the Warring States Period to the waning days of the Qin Dynasty and the dawning of the Han Dynasty. Notably, during the Warring States Period, the Yan state (燕国), driven by expansionist ambitions, encroached upon the northeastern frontiers, displacing numerous Indigenous tribes who sought sanctuary on the Korean Peninsula. Subsequently, the fall of the Six States under the dominion of the Qin Dynasty dealt a severe blow to the Qi state, an eastern powerhouse, compelling its populace to embark on a mass exodus. Further exacerbating matters, the turbulent twilight of the Qin Dynasty witnessed an insurrection, as chronicled in the "Records of the Three Kingdoms" (《三国志》), wherein multitudes from the Yan, Qi (齐), and Zhao (赵) regions sought refuge in the Korean Peninsula.

Compared to their ancient Jōmon predecessors, the Yayoi people gained substantial ground in weaponry, manufacturing techniques, and agricultural prowess, which swiftly propelled them to the forefront, surpassing the Jōmon populace in terms of sheer numbers. While some Jōmon individuals assimilated into Yayoi society, others valiantly battled and retreated, finding solace in the frosty confines of the northernmost Hokkaido isle.

Yet, among scholarly circles, an intriguing notion surfaces—a contention positing the Yayoi people as cultivators of rice, a crop indigenous to the southern regions of China. Consequently, the origins of the Yayoi populace may not be rooted in Northern China but rather find their genesis in the southernmost realms of the Middle Kingdom. Notably, investigations into the genetic makeup of Japan's two distinct rice strains have unearthed a peculiar revelation—one of these strains failed to traverse the Korean Peninsula, implying an alternate route of entry into the Japanese Archipelago. A plausible narrative emerges wherein ancient denizens of the Yangtze River Basin in Southern China, bearing the bountiful grains, embarked on a daring voyage, traversing the treacherous seas to arrive directly upon the shores of the Japanese isles. Thus, perhaps the ancestral lineage of the Yayoi people can be traced back to the primordial inhabitants of Southern China, forever entwined with the enigmatic tapestry of Japan's history.

This possibility lingers, yet upon careful consideration, the likelihood leans toward the Yayoi people hailing from the north. Take, for instance, the early Yayoi era in Japan, when the pivotal crop of wheat emerged, while simultaneously, the Chinese Yangtze River Basin remained devoid of its cultivation. Ergo, the introduction of wheat to Japan could only be traced back to the northern realms of China. Conversely, rice held a broader dominion in China prior to wheat's arrival. As we are aware, rice found its roots in the downstream region of the Yellow River approximately six to seven millennia ago, claiming its status as one of the three primary northern crops before the ingress of wheat.

Shandong, nestled in the lower realms of the Yellow River, boasted an enduring legacy of rice cultivation. When mapping out the course of Japanese crop importation, besides the Korean Peninsula, the Shandong Peninsula stood as another potential gateway. Hence, perchance Japanese rice emanated from the northerly expanses of Shandong, rather than the hallowed banks of the Yangtze River. Further conjecture whispers of a contingent of Yayoi people, embarking from the Shandong Peninsula, braving the high seas to reach their destined shores.

Fascinatingly, zoologists unearthed a revelation. Dogs and rats in the northeastern fringes of Japan were traced back to their origins and introduction from Southern Asia, during the heyday of the Jōmon Period or perchance even earlier. Conversely, the western reaches of Japan witnessed the influx of dogs and rats originating from Northern Asia during the Yayoi Period or possibly thereafter. Those noble beasts and scurrying rodents from the Jōmon era might have been borne by the hand of ancient predecessors, bestowed upon this land and left to weave their legacy. In contrast, the canine and rodent companions of the Yayoi era were unmistakably transported from the bountiful climes of Northern Asia by the Yayoi people themselves.

More compelling evidence arises from genetic research conducted in Korea. An exploration of 185 unrelated Koreans unveiled a remarkable revelation: 40% among them bear genes hailing from the populace inhabiting China's Central Plains. This disclosure signifies a substantial proportion of Korean lineage derives from the northern reaches of China. Furthermore, a trove of 25 ancient skeletal remains was exhumed from islands neighboring the Korean Peninsula, nestled alongside the Tsushima Strait. Scholars meticulously extracted their genetic code, illuminating a profound connection between these departed souls, modern-day Koreans, and the Japanese. Remarkably, these archaic individuals exhibited a fondness for ceramic vessels reminiscent of the Yayoi Culture. Weaving these enthralling strands of genetic inquiry together, one can glean a plausible notion: the progenitors of the Yayoi people likely embarked on a migration from the frigid realms of Northern China or the hinterlands of Northern Asia, employing the Korean Peninsula as a gateway to the storied Japanese Archipelago.

Beyond the prominent Yamato clan, Japan boasts two distinctive and sparsely populated ethnic groups—the Ainu and Ryukyuan people. Genetically akin to the Jōmon people, the Ainu find themselves among the lineage forcibly displaced by the Yayoi, banished to the remote reaches of Japan's northern territories. It is worth noting that throughout the epoch of 2,000 years, the Ainu did partake in certain genetic exchanges with their Yayoi counterparts. Similarly, the Ryukyuan people exhibit discernible genetic proximity to the Ainu, denoting their affiliation as one of the offspring spawned by the Jōmon lineage. Yet, the Ryukyuan people also share genetic affinities with the vibrant tapestry of Southeast Asian populations and the Indigenous ethnic communities of Taiwan. This suggests that during the epoch of expansion, the Austronesian linguistic groups influenced and permeated the very fabric of the Ryukyuan populace.

Henceforth, notwithstanding some ambiguities surrounding the introduction of rice, an overarching panorama emerges. The ancestral heritage of the Yayoi people evokes the ancient sages of Northern China. Furthermore, the more archaic Jōmon people trace their origins to the recesses of Northeast Asia, laboring under the isolation of their "isle" for countless millennia, evolving independently. Thus, the present-day denizens of Japan find themselves as the scions of a harmonious amalgamation—a captivating fusion—between the Yayoi and Jōmon lineage. Their forebears, in all likelihood, hailed from the ethereal realms of China's northern dominions, Northeast Asia, and the neighboring continent that dances with the waters.

8.4 The Ancient Bond Between Asian and American Lineages

"In the vast expanse that separates Asia from America, a slender strait alone connects them. The striking resemblance between Native Americans and their Eastern Asian counterparts fuels speculation, pondering whether the former are descendants of the latter or vice versa … "

In 1787, United States President Thomas Jefferson (1743–1826) penned his contemplations in the *Notes on the State of Virginia*, asserting a profound ancestral connection between Native Americans and Asians.

Jefferson's ruminations hold merit, for the Indigenous peoples of America not only share a bloodline with their Eastern Asian brethren but also possess an incontrovertible kinship with the ancient denizens of China.

Beyond their shared physical attributes, astute observers have long noticed a remarkable cultural affinity between Native Americans and the native inhabitants of Siberia. Unearthed in the 1950s at the Clovis site in New Mexico, a trove of archaeological artifacts unveiled an abundance of sharply honed stone implements, the famed "Clovis points." These stone spearheads found their resting place, many embedded within the skeletal remains of mammoths. Astonishingly, another site in Arizona revealed the surreal image of eight Clovis points artfully lodged within the carcass of a mammoth.

Yet, Native Americans were not the sole custodians of these formidable stone tools. In truth, the ancient Siberians wielded remarkably similar implements, akin to the Clovis points, which journeyed from Siberia to the vast American continent.

Radiocarbon dating lays bare the vivacity of the Clovis culture, flourishing approximately 11,000 years ago. However, the migration of ancient humans into the American continent likely preceded this epoch, estimated to have occurred some 15,000 years ago, or perhaps even earlier—a temporal parallel to the arrival of the ancestral Jōmon people upon the shores of the Japanese Archipelago, amidst the twilight of the Ice Age. During this frigid era, earth's icy grip transformed copious bodies of water into frozen expanses, resulting in a drastic plummet of sea levels. Consequently, Jefferson's description of the narrow strait—Bering Strait—assumes the form of a land bridge, steadfastly bridging the gap from approximately 34,000 to 11,000 years ago. Led by their nomadic instincts and the pursuit of sustenance, the intrepid ancient Siberians unwittingly traversed the thresholds of the American continent.

We know the prowess of the "steppe hunters" among modern *Homo sapiens*, skilled in the art of the hunt. They ventured eastward, traversing the vast Eurasian steppe, conquering the mountain passes of Tian Shan and Altai, and penetrating the realm of China. Descendants of the "steppe hunters" might well have inhabited those mountaintop caves. One can envision these adept hunters, residing for ages amidst the expansive Siberian grasslands and jungles. Those relentless hunters who pursued their quarry across the American continent are kin to the offspring of the "steppe hunters." Thus, from this vantage point, the ancient humans who set foot upon the American continent share a common progenitor with the dwellers of mountaintop caves. The Indigenous peoples of the Americas, the descendants of those ancient humans, also share ancestral ties with many modern-day Chinese. That shared ancestor, none other than the "steppe hunters."

In a time-honored site nestled within the southern reaches of Lake Baikal in Northern Asia, a revelation was unearthed: the fragmented dental remains of our ancient forebears, dating back 14,000 years. Scientists, through their diligent efforts, succeeded in extracting the genetic essence embedded within these teeth. Their analysis yielded an intriguing revelation—an undeniable genetic affinity between these ancient inhabitants and the Indigenous peoples of the Americas. Thus, a testament to their shared ancestral lineage. These forebears of yore, spanning the vast expanses of Northern Asia, held sway over territories far and wide. Some crossed the treacherous Bering Strait, evolving into the esteemed progenitors of the Indigenous peoples of the Americas.

Upon closer examination, the genetic makeup of Native Americans unveils itself as an exquisite tapestry, woven from the intricate threads of two ancient populations. One, hailing from the hallowed lands of Northern Asia, and the other, a remnant of the primordial denizens of Northeast Asia. Echoing through the ages, a unique genetic marker unites the ancient peoples of Northern Asia, dating back 14,000 years, with their brethren from Northeast Asia, who thrived nearly 10,000 years ago. This intermingling bespeaks of a profound exchange of genetic heritage during their shared pursuits of the bountiful hunt, bequeathing upon the migrating branches a fusion of the two lineages.

Delving into the realms of North America's Indigenous inhabitants, genetic scrutiny leads us to an astonishing revelation—ancestral roots trace back to a select few. A mere handful, numbering between ten and twenty individuals, stand as the progenitors of the great multitude populating the American continent today. Naturally, as the relentless march of time surged forth over 15,000 years, countless early ancients failed to bequeath their legacy to the present day. Thus, molecular biologists, taking into account this crucial factor, postulate a meager initial human presence in North America, ranging from a scant few dozen to several hundred intrepid souls. Hence, the genetic diversity among Native Americans remains notably subdued, paling in comparison to the amalgamation of populations on the sprawling Eurasian continent, and surely outshined by the mosaic of *Homo sapiens*' ancestral homeland—the great African continent.

Though their numbers were small, the intrepid trailblazers of the Americas exuded unwavering resolve and enduring valor. Across the treacherous expanse of the ice-cloaked Bering Strait, they forged ahead, undeterred by the formidable stretch that lay ahead. It was no mere jaunt. Emerging from the icy clutches, they ventured southward, delving into the very heart of the American continent, where the boundless North American prairies awaited, with open arms and untamed creatures aplenty, akin to mobile larders, tempting these audacious souls to feast to their hearts' content.

Armed with sharp-edged stone spearheads, brought forth from their Asian homelands, the primordial denizens of the Americas reveled in a blissful era of plentiful banquets. Within a mere millennium, they had traversed the American expanse, from the northern realms to the southern reaches. Three-quarters of the continent's mammalian brethren fell prey to their conquest, devoured in their entirety, including the esteemed equine brethren.

Eons ago, during the ancient epoch known as the Pliocene, a precursor to the modern equine, christened the "New World horse," roamed the North American lands. As time flowed, the evolutionary tide bestowed upon them the true equine form. Subsequently, these noble steeds embarked on a pilgrimage from North to South America, their influence extending even unto the Eurasian mainland, crossing the icy threshold of the Bering Strait. Indeed, the horses of Eurasia owe their origins to the American continent.

Regrettably, those horses that lingered in the Americas succumbed to extinction some ten to eight millennia ago, precisely during the epoch of grandiose expansion, as ancient humans zealously fanned across the continent. Thus, it is fair to deduce that the noble steeds of the Americas were prey to the ancestral hunts and culinary cravings of the Native Americans, echoing the manner in which the denizens of the Eurasian landmass, in their nascent days, turned to horses as sustenance.

Alas, not all horses upon the Eurasian terrain met such a dire fate, persevering through the agrarian and nomadic epochs, tamed to become stalwart companions of humanity. In the fifteenth century, Spanish colonizers introduced war horses to the American continent, bestowing upon several ancient civilizations a fateful and irrevocable blow. The people of that era, oblivious to their forefathers' encounters with these majestic, enigmatic beasts, simply consumed them, unaware of their profound significance.

Yet, the creature that befell a graver plight than the horse was the mammoth. The horse, at least, found solace upon the Eurasian canvas, heralding its triumphant era as the warhorse. However, whether on the Eurasian or American frontiers, the fate of the mammoth remained unchanging—they were all vanquished and consumed by ancient man, destined for complete annihilation.

Nonetheless, the absence of mammoths and true horses from the American continent is not solely the burden of the descendants of the "steppe hunters" that ventured forth. They likely do not stand alone as the sole progenitors of the Indigenous peoples of the Americas, nor do they hold the distinction of being the earliest inhabitants to set foot upon this captivating continent. Conceivably, the true pioneers of the Americas were the progeny of the Coastal Rampage among the modern *Homo sapiens*, forever etching their place in the annals of this land.

Chapter 2 of this book delves into the genetic juxtaposition of the Tianyuan Cave dwellers, who roamed the lands of ancient China some 40,000 years ago, and the diverse Indigenous tribes inhabiting the vast expanse of the American continent today. Scholars, with their keen interest, have embarked upon this comparative analysis and uncovered a fascinating discovery: the closest genetic bond can be observed between the denizens of the Amazon rainforest, thriving in the realms of modern-day South America, and those ancient Tianyuan Cave inhabitants. This revelation emphatically contradicts the notion of a singular migratory event shaping the Indigenous peoples of the Americas. Eons ago, it seems, a branch of modern humans, intimately linked to the Tianyuan Cave populace, migrated to the American continent and diffused throughout the lush foliage of the Amazon's southern reaches. As for the forebears of other native American tribes, their ancestral origins trace back to distinct migratory episodes, resulting in a comparatively distant genetic connection to the Tianyuan Cave people.

Yet, it appears the Tianyuan Cave dwellers may not be the sole distant relatives of the Amazonians' progenitors. Another remarkable genetic inquiry gathered DNA samples from nine Indigenous tribes in Brazil, juxtaposing them against the genetic makeup of 200 non-American Indigenous populations. Astonishingly, the findings unearthed an unexpected parallel: certain Amazonian tribes carry within them 2% of genes hailing from the Indigenous peoples of Oceania. Yes, indeed, tribes ensconced deep within the South American jungles bear genetic similarities to their brethren dwelling in the far-flung corners of Australia, New Guinea, and the Andaman Islands nestled within the Indian Ocean. Recall that the Indigenous peoples of Australia are the descendants of modern humans who departed Africa some 70,000 years ago. Dare we speculate that 10,000 to 20,000 years past, these seafaring souls embarked from Australia and New Guinea, spanning the entire Pacific Ocean to reach the shores of South America?

One must ponder the matter of the Indigenous peoples of Australia, marooned upon their isolated continent and a handful of remote islands for countless millennia. Can we truly ascribe to them the prowess to traverse the Pacific? Thus, the presence of Oceania's Indigenous genes within the Amazonians is explicable solely through the terrestrial bridge once gracing the Bering Strait. Hence, we may surmise that an ancient epoch witnessed the march of the Coastal Rampage, akin to both the Tianyuan Cave dwellers and the Indigenous peoples of Australia. This valiant band navigated the coastal edges of East Asia, forging onward into the northern realms of Asia, braving the icy expanse of the Bering Strait during the eons of glaciation, crossing over into the American continent, and resolutely forging ahead, ultimately reaching the verdant embrace of the Amazonian rainforest. Thus, it comes to pass that the Amazonian tribes bear within them the genetic imprints of both the Tianyuan Cave people and the Indigenous peoples of Australia, united by a shared forebear endowed with extraordinary seafaring prowess and migratory aptitude—the legendary Coastal Rampage.

Evidently, the descendants of these intrepid Coastal Rampage embarked upon the American continent long before their counterparts, the descendants of the "steppe hunters," and akin to the latter, they, too, participated in the relentless slaughter of majestic megafauna, inevitably bearing a measure of responsibility for the demise of the mighty mammoths.

Even in the annals of time, amid the vast expanse of millennia, sporadic migrations from the Asian continent to the American landmass occurred. Yet, the Indigenous peoples of the Americas before the era of Columbus stood as the least genetically diverse among all continents. Their genetic makeup derived from a meager fraction separated from the vast gene reservoir of Eurasia. Though they witnessed millennia on the American soil, evolving and undergoing genetic mutations, their genetic identity remained stifled in comparison to their kin on the Eurasian landmass, shackled by an overwhelming uniformity.

This spawned an immensely grave and latent menace, wherein a disease capable of inflicting grave health afflictions upon a tribe in the Americas could reverberate throughout the entirety of the Indigenous populace of the continent. Their genetic uniformity condemned them to a shared fate. The annals of history bear witness to this progression, as European colonizers, upon Columbus's discovery of the Americas, unleashed upon the land a maelstrom of afflictions that had long ravaged Eurasia. Smallpox, measles, and influenza surged through the veins of the American continent, laying waste to its Indigenous inhabitants. Scholars surmise that between the years 1492 and 1650, a plethora of epidemic outbreaks decimated the population, resulting in a loss ranging from 50% to 90%. Varying estimations exist among scholars. The European colonizers swiftly overpowered the Aztec Empire, the Inca Empire, and the confederations of the Maya, for firearms and horses paled in comparison to the true weapons of mass destruction—the pathogens of Eurasia.

Hence, the ascendancy of European colonizers over the native powers of the Americas lay not in technological prowess or the fallacy of cultural superiority, but rather in the intricate tapestry of their genetic code. It was not that the Spanish or Portuguese bore genes rendering them more sagacious than their Indigenous counterparts. They were all modern *Homo sapiens*, devoid of disparities in intellectual capacity. Instead, it was their genetic composition that fortified their resilience against the marauding pathogens.

The migratory streams flowing into the American continent did not flow eternally in one direction. They coursed back through the land bridge of the Bering Strait, tethering the Americas to Eurasia. Linguists have unearthed linguistic resemblances between the Indigenous languages of Central Siberia and those of North America's original inhabitants. Elements of language birthed in the Americas journeyed back to Siberia, encapsulating a temporal cycle. The nomadic wanderings of ancient humans were unbound by constraints, every cardinal direction beckoning as a destination. Those ancestral denizens of the Americas who retraced their steps to Eurasia surely melded into the tapestry of the prevailing population.

Be it the Polynesians dwelling upon Easter Island, isolated by unfathomable oceans, or the Indigenous tribes sheltered within the Amazonian rainforest, separated by half the globe, a tenuous genetic connection lingers between them and the Chinese people. In the realm of genetics, earth truly reveals itself as a grand "global village," its inhabitants, irrespective of their abodes, united as "villagers" within this tapestry. From the annals of antiquity to the present age, and beyond, this intricate network of connection and amalgamation persists.

The Column: The Enigma of Easter Island's Sweet Potato Origins

Chinese people bestow upon the humble sweet potato a plethora of names: "hongshu" (红薯), "ganshu" (甘薯), "baishu" (白薯), and "digua" (地瓜) to name a few. The rich nomenclature implies an intricate path of sweet potato dissemination into the realm of China. The fertile homeland of this tuberous marvel resides in the Americas, where its domestication took root thousands of sunlit harvests ago. It was during the era of audacious maritime endeavors that intrepid European voyagers propagated a panoply of American crops worldwide, sweet potatoes among them.

Amidst the eighteenth century, erudite minds stumbled upon an enigmatic tableau on Easter Island. Long before European mariners graced its shores, the Indigenous denizens had already woven sweet potatoes into the fabric of sustenance. Whence came this extraordinary botanical bounty to Easter Island's shores?

Contemplation ensues among scholars, veering toward the notion that ancient American inhabitants braved the treacherous seas, bearing with them the gift of domesticated sweet potatoes to this remote island paradise. Many moons prior, this bountiful crop had taken root, weaving its nourishing tendrils into the tapestry of the land. Yet, the Native Americans, skilled as they were in the terrestrial realm, seemed to possess scant nautical proclivities. Thus, another hypothesis emerges from the island's folklore: their forebears ventured across the ocean's expanse to the vast American continent, returning with the coveted spoils of precious sweet potatoes.

Irrespective of the veracity of either notion, both illuminate the realm of possibility that Easter Island and the American continent shared a historical narrative predating the Age of Exploration. This revelation casts an intriguing shadow of doubt upon the prevailing belief that the American diaspora solely transpired through the Bering Strait, prompting scholarly minds to dance with the enigma of transoceanic exchanges.

The study of sweet potato genetics has shed light on the enigma surrounding these tubers on Easter Island. Molecular biologists meticulously examined the

DNA samples of nearly 200 sweet potatoes, as well as more than 30 samples of their wild ancestral kin. Astonishingly, they traced the closest botanical relative of the sweet potato to the beautiful oca, resembling, in some small way, the oxalis flower. It appears that the sweet potato found its origins in the oca species as far back as 800,000 years ago.

Interestingly, the so-called Polynesian sweet potato varieties cultivated on Easter Island diverged from their American counterparts a staggering 100,000 years ago. This implies that well before modern humans wandered the African mainland, sweet potatoes had ventured beyond the Americas, making their way to the idyllic islands of the South Pacific, where they evolved independently into new cultivars. Thus, it becomes evident that the sweet potatoes on Easter Island were not borne by the hands of Native Americans or the islanders' ancestors but were delivered by the benevolence of nature. It is plausible that the seeds of American sweet potatoes were transported across the Pacific to these islands by hurricanes, ocean currents, or perhaps even the wings of birds, paving the way for their transformative journey into the Polynesian sweet potato.

In the latter half of the eighteenth century, the British seafarer, Captain James Cook, embarked on a grand expedition to the Southern Pacific, accompanied by sweet potatoes. However, it is worth noting that these particular tubers hailed from North America, distinct from their South American counterparts. Consequently, the sweet potatoes that presently thrive on the Pacific isles predominantly owe their existence to the North American varieties introduced by European explorers.

Yet, the Polynesian sweet potato is not the sole varietal testament to nature's extensive "delivery" service. Biologists, delving into the annals of time, have unearthed evidence of at least two other sweet potato species traversing the vast expanse of the Pacific Ocean. One found solace upon the shores of the Hawaiian Islands, while the other, in an astonishing feat, journeyed through the Polynesian Archipelago, crossed the Indian Ocean, and propagated itself upon the verdant lands of Madagascar in Southeast Africa.

Just as the tale of Easter Island's sweet potatoes seems to reach its denouement, linguists proffer an intriguing quandary. The astute scholars have stumbled upon a linguistic convergence—the Polynesian term for "tuber" shares a striking phonetic resemblance with the native Quechua tongue spoken by the Indigenous peoples of the majestic Andes in South America. Surely, this cannot be dismissed as mere happenstance.

In the ancient history, did the American continent and Easter Island, in truth, forge connections, enabling the exchange of knowledge and life's bounties?

Emerging from Isolation, the Union of a Family Across the Vast Seas

Genes flow ceaselessly, coursing through the veins of ancestors and descending upon you and me.

Many genes of yore still find solace within our cells, while each generation, in its evolutionary journey, forges fresh genes through the alchemy of mutation. The human genome, a tapestry of inheritance and progress, has been woven across epochs, tides, and cardinal points.

Should you query, "Who were our progenitors?" I shall counter, "Pray tell, which time and place stir your curiosity?"

Ancestors, the sum of bygone generations, dwelled within specific chronicles. Contemplating the ancestral legacy of the Chinese, its procreation and migratory odyssey, we may parse the countless eons into tiers, employing a "stratified ancestry" mentality (as delineated in the chart).

The primordial stratum harks back to the era of our zeroth ancestors—those upright beings, *Homo erectus*. From the heartlands of Africa, these earliest *Homo erectus* embarked on an epic pilgrimage across the vast expanse of West Asia and Central Asia. By the time they reached the shores of East Asia, they likely embodied the twilight phase of *Homo erectus*, quite possibly venturing into China's embrace from its northern realm. Among them, the Peking Man of Zhoukoudian emerges as a quintessential exemplar, an emblem of China's *Homo erectus* lineage, perchance even among the primordial ancestors of China. While not our immediate forebears, their arrival, in and of itself, bears testament to their resilience—piercing the geographical barriers of Asia's rugged terrain and waterways. From the very inception of China's sacred soil, a profound kinship was forged, bridging the divide with the world at large, even if these connections ebbed and flowed like torrents or thread-thin whispers.

Ancestral layers

Layer of our ancestry	Period	Drivers to migration	Tendency of migration
Fourth Ancestral Layer	From 4,000 years ago to the present	Civilization, Culture	Interaction between Eastern and western populations, with Northern populations migrating southward
Third Ancestral Layer	From 10,000 to 4,000 years ago	Agriculture	Expansion of farming communities
Second Ancestral Layer	From 19,000 to 10,000 years ago	Fishing and hunting gathering	Southern populations moving northward
Period of contraction	From 26,500 to 19,000 years ago	Harsh cold climate	Population retreating to the south
First Ancestral Layer	From 50,000 to 26,500 years ago	Fishing and hunting gathering	*Homo sapiens* entering East Asia from both northern and southern directions
Zeroth Ancestral Layer	Ancient times to 50,000 years ago	Hunting and gathering	*Homo erectus* entering East Asia from the western direction

In the days of East Asian *Homo erectus*, when genetic and technological exchanges between the East and West of the Eurasian continent were as delicate as silk threads, it seemed as if the population of East Asian archaic humans found themselves stranded on a colossal island, cut off from the outside world, particularly the distant lands of Western Eurasia and Africa. Their communication was meager, relying primarily on the rudimentary stone tools they had brought with them. Over the course of hundreds of thousands of years, they persevered, living in solitude, their existence a struggle against the tides of time. Eventually, they vanished, leaving behind an interrupted river of their genes. It is this absence of genetic legacy that prompts me to designate them as the "Zeroth Ancestral Layer."

From 50 millennia ago, East Asian ancient humans embarked upon the first ancestral layer, an epoch marked by the expansion of *Homo sapiens*. This march of *Homo sapiens* in East Asia began in the southern territories before an interplay between the ancestral *Homo sapiens* of the southern and northern regions orchestrated a migratory spectacle akin to a seesaw. China, the vast expanse, resonated with vibrancy. The wisdom of *Homo sapiens*, both individually and collectively, endowed them with an extraordinary capacity to traverse diverse landscapes—be it the expansive seas or majestic mountains—armed with the requisite skills and unwavering courage to defy the spell of isolation.

During the era of *Homo sapiens*' expansion, geographical barriers on the mainland crumbled, surrendering to the void that pervaded the plains, jungles, mountains, and even the coastal islands. The ancient forebears, in their migration, enacted a symphony of filling the void, and the genetic river of *Homo sapiens* surged forth, inundating every corner of Earth suitable for habitation. On the Chinese terrain, their footprints traced along the coastlines, surging forth from the tropical South China Sea, northward to the temperate embrace of Bohai Bay, before venturing deep into the mainland. Alternatively, they followed the untamed beasts' tracks across the steppe, sweeping eastward from the western frontiers, engulfing the northern hillocks and sprawling plains ...

Amidst the relentless onslaught of the Last Glacial Maximum, spanning from 26,500 years ago to the distant horizon of 19,000 years ago, our ancient ancestors endured a perilous phase of regression. The northerly ancient populace suffered grim losses, while a few might have sought solace in a southern retreat. The southern brethren, harnessing the benefits bestowed by the ebb of sea levels, capitalized on the merging of continents and islands, fostering a sanctuary of respite to nurture vitality and cultivate resilience.

After surviving the Last Glacial Maximum, the second layer of our ancestors emerged. From their slumber in the southern depths, where they had lain dormant for millennia, they seized the opportune moment of a warmer interglacial period. With spears aloft, they ventured northward, driven by the chance to pursue thriving wild creatures and gather the ripe fruits and untamed greens that adorned the mountains. The ancient humans of the second layer migrated predominantly from south to north, and this expansion etched its mark upon their genetic fabric, endowing the present-day populations of the south with a richer genetic tapestry than their northern counterparts.

From the primeval stage of the zeroth layer of ancestors to the first layer and onward, our forebears subsisted primarily through the art of hunting, gathering, and fishing. Such survival skills propelled them on a relentless odyssey, traversing the vast expanse of lands. Then, some 10,000 years ago, they stepped into the realm of the third ancestral layer, embarking upon a seismic transformation—ushering in the era of agriculture.

The agricultural epoch in East Asia unfolded in an enchanting fashion. Whether north or south, the fabled land of China seemed imbued with an enchanting abundance, as if touched by the wand of some mystical sorcerer. Fields teemed with a cornucopia of crops, and domesticated creatures of all kinds flourished. And the master sorcerers behind this marvel? None other than our wise ancestors—the pioneering cultivators of the land. Amidst the "darkness before dawn" of the Last Glacial Maximum, they ardently experimented with the taming of plants and animals, their efforts kindling a revolutionary metamorphosis come the waning of the Ice Age.

The agricultural era also heralded a population boom, its fertility entwined with the exponential growth of grain harvests. Across the sprawling plains, the denizens of the Middle Kingdom encountered, for the first time, the claustrophobic embrace of density. Wave upon wave of humanity embarked upon their journeys, clutching in their hands the seeds of millet, maize, or rice, accompanied by their faithful companions of swine, canines, fowl, and waterfowl. They sought new frontiers ripe for cultivation, transforming the four corners of the land into an enchanting stage for the magic of agriculture.

Relative to their hunter-gatherer brethren, the agricultural populace swelled in overwhelming numbers, an indisputable dominance that resonated in the array of genetic types. From the vantage point of genetic expansion, this epoch witnessed the migration of communities in the grand tableau of the "Replacement Model," particularly the substitution of agricultural groups for their hunting and gathering counterparts, all within the fold of the modern *Homo sapiens*. Thus, the dissemination of agriculture signified not only the widespread propagation of agricultural genetic strains but also the absorption of minute traces of genetic heritage from dwindling hunting and gathering enclaves.

From the ancestors who roamed the Earth some 4,000 years ago, we trace the lineage of our more recent forebears, nestled snugly within the fourth ancestral layer, an era defined by the propagation of cultural genes. The ethereal melodies of civilization resounded as the East Asian realm gracefully embraced its cultural genetic epoch. Firmly rooted in the bountiful soils of the Middle Kingdom, our progenitors embraced a rich tapestry of resources and technologies flowing from the western reaches of the continent. It was the allure of bronze craftsmanship and the advent of horse-drawn carriages that proved vital in erecting and sustaining the very fabric of the nation's institutional framework. Innovating upon these foundations, they orchestrated a harmonious symphony, heralding the dawn of China's unique ancient civilization.

This prelude, however, was far from a solitary recital confined to the Central Plains. It was an ensemble performed in unison by the manifold civilizations that thrived across the vast expanse of the north and south, as if the grand rivers themselves provided an ethereal stage. While each dominion boasted a distinct ethnic core, it was not the biological essence that set them apart, but rather the cultural essence that coursed through their collective veins. Within this realm, where ethnic convergence was a frequent and unabating affair, it was cultural genetic codes that truly defined the boundaries between realms. Communities forged through the amalgamation of biologically diverse groups found unity in shared cultural identities, giving rise to tribes, clans, and formidable states. The interplay and rivalry between these entities transcended the realm of mere biological exchanges, becoming a dance of cultural genes, intertwining and colliding in exquisite complexity.

In the early epochs of the fourth ancestral layer, it was the vast grasslands of the west that bestowed upon us the blessings of vital technologies, imparting a chronological advantage. Thus, the flux and flow of biological and cultural genes, akin to a celestial choreography, gravitated toward the east. But once the cultural tapestry of East Asia absorbed the offerings from the western realms and wove them into its own vibrant fabric, a metamorphosis occurred. The spirit of expansion, symbolized by the Qin and Han dynasties in the heartland and the indomitable *Xiongnu* on the Eastern steppes, charted a path from east to west, carrying with it the very essence of East Asian peoples, both in their biological and cultural forms, into the recesses of the northwestern frontiers.

Simultaneously, the resolute Han populace of the northern realms, alongside various regional groups such as the ancestral forebears of the Austronesian-speaking tribes, embarked upon a resplendent odyssey. Fueled by the vigor of a thriving agrarian populace, they ventured forth, their collective essence cascading like golden embers. With unwavering determination, they forged southward, scattering their biological and cultural genes across the fertile landscapes of the south, permeating the realms of East Asia and the islands strewn across the Southeast Asian archipelago. Their intrepid spirit, carried on the crest of mighty waves, even dared to traverse the vast and enigmatic depths of the boundless oceans that lie beneath the celestial tapestry.

The above serves as a coarse recollection of the ancestral layers entrenched within the vast river of history. It encapsulates the essence of the "unwritten chronicles" deciphered by scholars from various fields, extracted meticulously from the annals of our progenitors and the genetic tapestry of flora and fauna. Perusing this untold saga transcribed in the language of genes, we unearth the grandeur and undulating epic of our ancestors' migrations across the expanse of the Middle Kingdom. As descendants, what revelations can we glean from this celestial opus?

Primarily, we are not solitary heirs of East Asia, but the product of confluence —our forebears and our civilization intertwined.

Though some scholars persist in championing the autochthonous origins of the Chinese lineage or the purebred genesis of Chinese civilization, evidence drawn from ancient gene comparisons among diverse ethnicities and the genetic exploration of the animal and plant realms intimates that our progenitors first hailed from Africa. Subsequently, a tapestry of ancient peoples from all corners amalgamated, birthing the Chinese lineage. The tapestry of Chinese civilization, a testament to the toil and sagacity of our ancestors, resulted not only from their assiduous cultivation on the fertile soil of East Asia but also from the rich infusion of external resources, technology, and a myriad of ideas that took root and bore fruit. The demarcations we recognize today, delineating continents and nation-states, are but recent artificial boundaries forged within the last few centuries—absent in the path of ancient migrations. The ancients did not confine themselves to the confines of modern borders, limiting their genetic and cultural exchanges.

Secondly, genes converge upon a shared destiny, and culture harmoniously strides toward unity.

The quantity and density of populations form the crux of genetic interaction among the ancients. In the age of *Homo erectus*, populations were scarce, and tribes modest in size. Many tribes perished amidst the ebb and flow of natural vicissitudes, their might scarcely surpassing the primates that surrounded them—gorillas and troops of monkeys. Consequently, distant genetic exchanges languished, while *Homo erectus* populations scattered across the continents evolved independently. The notion that *Homo erectus* populations on disparate continents would metamorphose into modern *Homo sapiens* in synchrony is hardly factual.

As modern *Homo sapiens* ventured beyond the bounds of Africa, disseminating throughout the world, their numbers burgeoned exponentially. Their remarkable migratory prowess kindled intermingling with Neanderthals and Denisovans along the path, not to mention the interbreeding of various *Homo sapiens* groups. The advent of the agricultural epoch witnessed a surge in population, surpassing severalfold the hunter-gatherer era of yore. Genetic exchanges between diverse ethnic groups became ever more frequent, engendering a gradual assimilation of genetic traits among disparate peoples. In due course, humanity's genetic tapestry would converge upon a shared destiny. While distinctions may endure between groups, they reside primarily within the realms of cultural heritage, rather than biological inheritance.

In the end, genes reside within the realm of life sciences, while ancestors reside within the realm of cultural sentiment. Along the path to understanding our forebears and our own selves, both reason and emotion are indispensable.

Science, a bastion of rationality and sagacity, remains indifferent to our inclinations, presenting its case through irrefutable evidence and uncovering the clandestine history of our progenitors. Without the scaffolding of scientific achievements, our very own lineage would be shrouded in perplexity. Nothing besmirches our ancestors more than misidentifying their true essence, worshipping false idols. To grasp unequivocally the identity of our direct ancestors, the life sciences stand as the paramount tool, for within the recesses of the cellular domain, etched upon the indiscernible genes, lie the vital signs of our lineage, uniting us irrevocably with our forebears.

The Chinese have long revered their ancestors, a reverence embedded within the ancient glyphs of the Shang and Zhou dynasties, which brim with tributes to those who came before. One can envision that well before those dynasties, denizens of the Middle Kingdom had already kindled a deep-seated veneration for their progenitors, establishing a sacred bond through worshipful rituals. This practice not only fortified ancient souls against a perilous external world but also imparted a sense of anchorage to the humble self, enabling the minuscule individual to grasp their place amidst the ceaseless ebb and flow of history. Thus, they acquired a palpable sense of existence and purpose.

The profound tradition of ancestral worship, ensconced within the annals of Chinese civilization and extending to the wider domain of East Asia, traces its origins to the archaic agrarian communities of yore. It manifests as a distinctive cultural strand within the Chinese genetic tapestry, diverging from the religious customs prevalent in the ancient civilizations of Europe and America. Even in the present era, the reverence and ceremonial homage to ancestors remain as vital expressions of sentiment for countless Chinese, an ethereal sanctuary they cannot sever from their lives and beings.

Confronting our progenitors, we embrace the weight of genetic evidence, while cherishing our spiritual heritage.

However, we must oppose any form of nationalism and ethnic discrimination built on the pretext of genetics, as these arguments lack scientific foundation.

All human beings on this planet are modern *Homo sapiens*, tracing their lineage back to a shared ancestor known as Mitochondrial Eve from 140,000 years ago and a common forefather named Y-chromosomal Adam from 60,000 years ago. For contemporary Chinese individuals, the common ancestors of their lineage would not predate 30,000 years. Tracing the origins of the majority of Chinese people may simply require turning back the hands of time by a few millennia to uncover success.

Keen readers might observe that this book endeavors to eschew terms like "Caucasian race," "Mongoloid race," "Aryan race," or "Negroid race," which possess racial undertones or even discriminatory implications, unless necessary to reference the viewpoints of early scholars. These terms blur the boundaries between modern ethnic concepts and ancient population designations. Often, they are concocted notions lacking genetic support, causing disturbance to our comprehension of human history. In this book, when depicting ancient populations, neutral geographical appellations pertaining to continents and cardinal directions are employed to denote these groups, such as "East Asian populations" or "people of the Western Steppe." On occasion, phrases incorporating ethnic names, like "Di-Qiang ethnic group" or "Xizango-Burman language group ethnic group," are utilized. These expressions have long been employed by scholars either as descriptions of ancient populations, avoiding confusion between ancient and present eras, or as neutral linguistic descriptions refraining from invoking racial discrimination.

It is imperative to emphasize that the notion of ethnicity we currently discuss emerged only with the advent of modern nation-states. It is an imagined community, a byproduct of cultural genetics. Genetic evidence reveals that as early as 10,000 years ago, during the agricultural epoch, the biological demarcations of population groups began to blur, facilitating a growing frequency of genetic exchange between various communities. Attempting to retroactively extrapolate ancient times using the cultural concept of "ethnicity" conceived by scholars centuries ago and asserting that clear genetic barriers and fundamental distinctions existed between numerous ethnic groups throughout the millennia-spanning migration history of humankind, subsequently determining intellectual and cultural hierarchies among these groups, represents a baseless conjecture devoid of scientific substantiation.

For instance, throughout history, Han Chinese people developed their own sense of ethnic identity, as did the Xiongnu people and the Qiang people. This identification was fortified by defining and venerating their respective ancestors and deities. It is an emotional connection forged through historical circumstances, rather than one with substantial genetic underpinnings. The essential disparity between nationalism and ethnic sentiment resides in the former propagating external discrimination while the latter seeks self-appreciation.

Within an ethnic enclave, the indulgence of self-appreciation poses no harm or offense, as long as it refrains from descending into external discrimination. Yet, we must remain vigilant of the lurking perils that accompany such self-indulgence—an insidious slippage toward the "Tragedy of Tasmania."

In the annals of Easter Island's folklore, amidst the initial influx of immigrants, resided a singular architect of note, *Nuku*. He alone possessed the skills to erect dwellings, while the rest of the migrants failed to grasp the architectural essence, particularly the art of roofing. Thus, upon Nuku's demise, the island's abodes proved defenseless against the whims of tempestuous winds, relinquishing their roofs to untamed gusts. As the voyage to Easter Island commenced, Nuku, regrettably, left his wife behind, rendering the immigrants incapable of retracing their steps to their ancestral abode. Henceforth, when dusk descended, a somber veil draped over the heart of the architect, for the setting sun cast its glow upon the direction of his wife and the cherished homeland.

Throughout ancient times, geographical barriers wreaked havoc upon Easter Island, Tasmanian Island, the Japanese archipelago, the Australian continent, and even the vast lands of China and the Americas, entangling ancient civilizations and ethnic communities in the web of the "Tragedy of Tasmania." In this era, blessed by the twenty-first-century technological marvels, we need not harbor concerns regarding the geographic divisions imposed by nature, estranging the great land of China from distant realms. However, we must exercise caution, for the seeds of obstructions, nurtured by the self-adulation prevalent within ethnic enclaves, bear the potential to germinate anew, erecting barriers between groups.

Take, for instance, the aboriginal populations that take pride in their organic evolution and the pristine origins of their indigenous cultures yet shamefully recoil from the assimilation of alien genetic material and cultural influences. Such an insular "emotional infatuation" fixates upon the dogma of pure lineage, manifesting as cultural haughtiness, artificially obstructing the interplay between native enclaves, nations, and the wider world.

Another cultural gene that ensnares a collective into a state of seclusion is the superstition of the jungle's law, or as it is often dubbed, the "Dark forest hypothesis."[1] Its adherents firmly assert that others are nothing short of hell, deeming foreign cultures and civilizations as their inevitable adversaries. To safeguard their own existence, they advocate severing ties with any form of engagement with other cultures and civilizations. Granted, throughout history, there have been instances where the phenomenon of the jungle's law has held sway—such as the Han Dynasty's impact on feeble cultures in the northwest region, or the injury

1 The dark forest hypothesis is the conjecture that many alien civilizations exist throughout the universe, but they are both silent and paranoid. In this framing, it is presumed that any space-faring civilization would view any other intelligent life as an inevitable threat, and thus destroy any nascent life that makes itself known. As a result, the electromagnetic spectrum would be relatively quiescent, without evidence of any intelligent alien life, as in a "dark forest" filled with "armed hunter(s) stalking through the trees like ghosts." The hypothesis was described by astronomer and author David Brin in his 1983 summary of the arguments for and against the Fermi paradox.
—Translator's note.

and disintegration inflicted upon the Native American civilization by European colonizers. Yet, the human saga of migration is primarily one of openness rather than confinement. It was the inadvertent seclusion of East Asia's *Homo erectus* that led to a stagnation in the realm of stone tool technology. On the contrary, the Central Plains' myriad cultures, driven by proactive openness, absorbed copious amounts of foreign wars, technologies, and even ideologies, thus ascending to the center stage of ancient civilization ... As the post-agricultural era unfolded, with burgeoning populations and the amalgamation of ethnicities, the dark forest law gradually lost its efficacy, for the more enclosed one became, the further they regressed, unable to impede the forces that rend asunder the barriers of seclusion.

Once one rebuffs communion with the outer realm, each person metamorphoses into an isle unto themselves, rendering every ethnic group, every nation an isolated island. Throughout the expanse of time, humanity, through genetic mutation and expansion, as well as the genesis and fusion of cultural genes, ultimately transcended the solitudes of the individual and the ethnic enclave, evading the poignant calamity of the Tower Isle. It behooves us, then, to embrace those cultural genes that foster the equitable convergence of diverse ethnic groups while repudiating those that cleave asunder the fabric of unity, engendering discriminatory hierarchies amidst the human tribes.

This book was penned from the close of 2019 to the dawn of 2020, catching the tumultuous period of the outbreak of the COVID-19 pandemic. The contagion coerced nations into shuttering their borders, while communities restricted the ingress and egress of individuals. It is a poignant testament to witness a virus, with its infectious tendrils, traversing all corners of the globe, accentuating the infinitesimal disparities in our biological essence. The diverse ethnic groups, separated by vast geographical expanses, had long been united through the intricate intermingling of their genetic tapestry, albeit revealed through the crucible of affliction.

In the throes of this crisis, enlightened minds across the globe implored for global collaboration, a collective resistance against the viral onslaught. Those forced into temporary seclusion harnessed the power of the internet, orchestrating events like the harmonious opus titled "One World: Together at Home," an offering of solace to a beleaguered world, illuminating the path of hope. In a spirit of intellectual camaraderie, scientists from every nation shared insights into the virus itself and their endeavors in the pursuit of antiviral remedies. It is through the tenacity of our ancestral heritage, a lineage perpetuated through the annals of history, that this active and open cultural genetic code manifests, bestowing upon us a fertile landscape of shared humanity.

The creed of "One World, One Family" is inscribed within the very fabric of our biological essence, passed down through the generations, beckoning us to etch it indelibly into the annals of our cultural heritage.

As mankind has scaled the lunar heights, our expedition beyond the confines of the Earth-Moon system remains elusive. From this vantage point, we are akin to solitary denizens adrift upon an insular isle within the vast cosmic seas. Just as our forebears, over countless millennia, cast their gaze upon the Earth's surface in ceaseless exploration, our pursuit of celestial inquiry shall endure unabated. There will come a day when intrepid astronauts shall break free from the terrestrial embrace, alighting upon the hallowed grounds of Mars and celestial spheres beyond. They shall inscribe a new chapter, an epic saga in the annals of human genetic expansion, unfaltering in their valor and immortal in their legacy.

Such will be the resplendence of Earth's life essence, and such shall be the collective pride of all humanity.

www.ingramcontent.com/pod-product-compliance
Ingram Content Group UK Ltd.
Pitfield, Milton Keynes, MK11 3LW, UK
UKHW021838100926
8556IPUK00004B/407

9 783034 360500